Brown/Foote/Iverson

2003 Update with MCAT for

Organic Chemistry

William H. Brown
Beloit College

Christopher S. Foote
University of California, Los Angeles

Brent L. Iverson
University of Texas, Austin

Australia • Canada • Mexico • Singapore • Spain • United Kingdom • United States

Spectra courtesy of Pouchet and Behnke, *The Aldrich Library of 13C and 1H FT NMR Spectra*, Vol. 1, 1993.

COPYRIGHT © 2004 Brooks/Cole, a division of Thomson Learning, Inc. Thomson Learning™ is a trademark used herein under license.

Information reprinted with permission from:
MCAT Sample Exams, ©2002 Stephan Bosworth, Marion A. Brisk, Ronald P. Drucker, Edgar M. Schnebel, Denise Garland, and Rosie M. Soy.
MCAT Success 2003, ©2003 Stephan Bosworth, Ronald P. Drucker, Edgar M. Schnebel, Denise Garland, Rosie M. Soy, and Marion A. Brisk.

ALL RIGHTS RESERVED. No part of this work covered by the copyright hereon may be reproduced
or used in any form or by any means—graphic, electronic, or mechanical, including but not limited to photocopying, recording, taping, Web distribution, information networks, or information storage and retrieval systems—without the written permission of the publisher.

Printed in the United States of America
1 2 3 4 5 6 7   07 06 05 04 03

Printer: Patterson Printing Company

ISBN 0-534-46588-9

For more information about our products, contact us at:
Thomson Learning Academic Resource Center
1-800-423-0563

For permission to use material from this text, contact us by:
Phone: 1-800-730-2214
Fax: 1-800-731-2215
Web: http://www.thomsonrights.com

Asia
Thomson Learning
5 Shenton Way #01-01
UIC Building
Singapore 068808

Australia/New Zealand
Thomson Learning
102 Dodds Street
Southbank, Victoria 3006
Australia

Canada
Nelson
1120 Birchmount Road
Toronto, Ontario M1K 5G4
Canada

Europe/Middle East/South Africa
Thomson Learning
High Holborn House
50/51 Bedford Row
London WC1R 4LR
United Kingdom

Latin America
Thomson Learning
Seneca, 53
Colonia Polanco
11560 Mexico D.F.
Mexico

Spain/Portugal
Paraninfo
Calle/Magallanes, 25
28015 Madrid, Spain

Table of Contents

NMR Analysis 3

MCAT Sample Questions 21

MCAT Answers and Explanations 67

Organic OWL 97

ANALYSIS OF NMR SPECTRA

"Things should be made as simple as possible, but not any simpler."

Albert Einstein

1.1 Review of Basic NMR Concepts

NMR analysis is a complex topic, from both the spectral interpretation and fundamental physics points of view. This section supplements the text on both of these fronts by first providing a concise summary of important general material, followed by the background necessary to interpret ^{1}H-NMR spectra for molecules that have more complex signal splitting patterns. We assume that you are already familiar with the material in the text.

A. The Classic ^{1}H-NMR Experiment

In the classic ^{1}H-NMR experiment, a sample is placed in a strong magnetic field. The absorption of electromagnetic radiation is measured as different ^{1}H nuclei are excited from their +1/2 spin states to their –1/2 spin states, a process referred to as **resonance**. The +1/2 nuclear spin state is the lower energy spin state in which the ^{1}H nuclear spin is aligned *with* the strong external magnetic field. The –1/2 nuclear spin state is the other "allowed" nuclear spin state, and corresponds to the higher energy situation in which the nuclear spin is aligned *against* the external magnetic field. You should refer to Figure 13.1 of the text.

There are two fundamental physical principles that are essential for understanding NMR as it relates to the determination of molecular structure.

1. The energy difference between +1/2 and -1/2 nuclear spin states is proportional to the strength of the magnetic field at the nucleus (Figure 13.2 of the text). This means that electromagnetic radiation of higher energy is required for resonance in a stronger magnetic field, and *vice versa*.

2. The electron density around a nucleus is induced to circulate by the external magnetic field, and the induced electronic circulation sets up it own small magnetic field that directly *opposes* the strong external magnetic field. This induced magnetic field effectively **shields** the underlying nucleus a small but measurable amount from the external magnetic field.

Combining these two fundamental principles, the more electron density around a nucleus, the greater the shielding from the external magnetic field. The greater the shielding, the smaller the net magnetic field present at the nucleus, so the lower the energy of electromagnetic radiation required to bring that nucleus into resonance, in other words, "flip its spin." Energy is proportional to electromagnetic radiation frequency, and resonance frequencies are plotted on an NMR spectrum. *Putting all of these ideas together, a nucleus that is surrounded by greater electron density will generally come into resonance at lower frequency than a nucleus surrounded by less electron density.*

To make spectra comparable from machines of different magnetic field strength, the frequency of resonance is plotted as a **chemical shift**. A chemical shift is the resonance frequency of a particular nucleus compared to that of a standard molecule, TMS in ^{1}H-NMR, scaled to the magnetic field strength of the NMR spectrometer and reported as parts per million, **ppm**. The units here are parts per million because the chemical shift changes associated with electron density differences are about one millionth as large as the external magnetic field used in an NMR spectrometer.

Hybridization state and the nature of any adjacent pi bonding electron density also influences chemical shift. Hydrogen atoms attached to aromatic rings are influenced the most due to the **induced ring current** of the aromatic pi electrons. H atoms attached to aromatic rings appear in the 6.5 ppm to 8.5 ppm range. Vinyl H atoms, i.e. those attached to alkenes, appear between 4.6 ppm and 5.7 ppm.

In a ^{1}H-NMR spectrum, a signal with a chemical shift that is larger, for example 4.8 ppm, corresponds to an H atom that either has relatively little electron density around it (it is adjacent to electronegative atoms or electron withdrawing groups) or is one that is attached to a carbon atom taking part in a pi bond. A signal with a smaller chemical shift, for example 1.8 ppm, is due to an H atom that has greater electron density around it, indicating it is not adjacent to any electronegative atoms, electron withdrawing groups, or pi bonds.

Different functional groups have different characteristic chemical shifts, so *comparing a given signal in an NMR spectrum to a reference table of chemical shifts (Table 13.2 in the text) allows the identification of functional groups in a molecule.*

B. Signal Splitting – The Short Version

In ^{1}H-NMR spectra, signals are split into several peaks by other H atoms that are not more than 3 bonds away. The signal splitting is caused by **spin-spin coupling** between adjacent nuclei. *Signal splitting allows the determination of how different functional groups are connected in a molecule, because atoms of only adjacent functional groups can split each other.* The general rule is that a signal will be split into **n + 1** peaks if there are **n** H atoms three bonds away. For example, the signal for an H atom with two H atoms that are three bonds away will be split into 2 + 1 = 3 peaks. The three peaks will appear in relative size ratios of 1:2:1, in accord with **Pascal's triangle** (Figure 13.12 in the text). Three adjacent H atoms would split a signal into 4 peaks of relative ratios 1:3:3:1, and so on.

Two H atoms on the *same* C atom will not split each other if that C atom is freely rotating and sp^3 hybridized, because the rapid bond rotation means that, on average, the two H atoms see the same chemical environment. These two H atoms are said to be **equivalent** and give rise to only one signal in an NMR spectrum, with an area that **integrates** to a relative value of 2 H atoms. Recall that in ^{1}H-NMR spectra, relative signal integrations (i.e. the area under the signals) are proportional to the number of equivalent H atoms giving rise to that signal. The three H atoms on a freely rotating –CH$_3$ group are also equivalent and their signal will integrate to a relative value of 3 H atoms.

For example, the two equivalent H atoms of a –CH$_2$- bonded to a –CH$_2$- on one side and a –CH$_3$ on the other has a total of 5 H atoms that are three bonds away. The signal for this central –CH$_2$- group is therefore predicted to be split into 5 + 1 = 6 peaks and will integrate to a relative value of 2 H atoms. The 100 MHz spectrum for 1-chloropropane, Cl–CH$_2$-CH$_2$-CH$_3$, is shown in Figure 1. The central –CH$_2$- of 1-chloropropane, labeled as "b" in Figure 1, is indeed split into 6 peaks and integrates to a relative value of 2 H atoms. Similarly, the signal for the –CH$_2$- bonded to Cl, labeled as "a" in Figure 1, is spit into 2 + 1 = 3 peaks and integrates to 2 H atoms. The signal for the –CH$_3$, labeled as "c", is split into 2 + 1 = 3 peaks and integrates to 3 H atoms. The signals are expanded and plotted as insets on the spectrum to clarify the splitting patterns. The integral is indicated by the continuous line plotted on the spectrum above the base line, and the relative rise of the integral is proportional to the relative number of H atoms associated with a signal.

The signal for the –CH$_2$- group attached to the electronegative Cl atom, labeled as "a", appears at a higher chemical shift (referred to as being **downfield**) compared to the other signals. There is less electron density around the "a" H atoms because they are closest to the electronegative Cl atom. Recall from the discussion above that

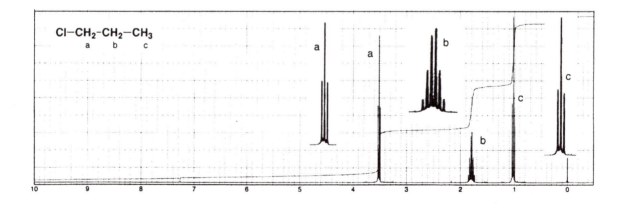

Figure 1 100 MHz ^{1}H-NMR spectrum of 1-chloropropane.

less electron density means *less* shielding and therefore a slightly *stronger* magnetic field at the nucleus and a *higher* frequency required for resonance.

Symmetry can make atoms equivalent in molecules. If atoms or groups are in identical locations relative to a **plane of symmetry** or **center of symmetry** in a molecule, then these atoms or groups are equivalent. For example, the two methyl groups of 2-chloropropane are in identical locations with respect to the plane of symmetry that bisects the molecule, so these methyl groups are equivalent and all six methyl group H atoms are equivalent. This means these six H atoms are in the same chemical environment. These atoms give rise to a single signal, integrating to 6 H atoms, that is split into 1 + 1 = 2 peaks by the central H atom. The signal of the central H atom is split into 6 + 1 = 7 peaks (the two peaks on the ends are small but discernable) and integrates to 1 H atom.

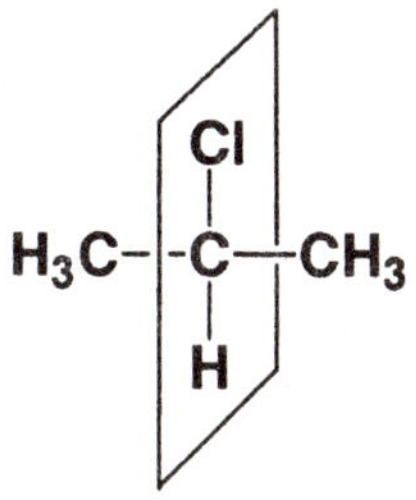

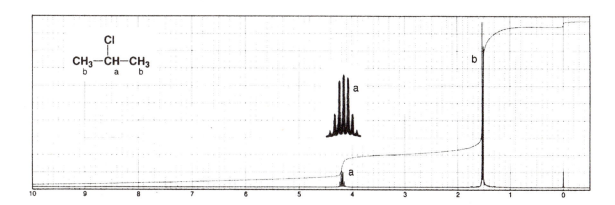

Figure 2 Structure of 2-chloropropane showing the plane of symmetry responsible for making the two methyl groups, and therefore the six methyl group H atoms, equivalent.

Figure 3 100 MHz ^{1}H-NMR spectrum of 2-chloropropane.

C. General Method

The only way to get proficient at molecular structure determination using NMR is to practice. There are numerous problems in the text at the end of Chapter 13, as well as subsequent chapters. A general method for analyzing NMR spectra can be described as follows:

1. Compare signal chemical shifts with reference tables to determine which functional groups may be present.

2. Evaluate signal integrations to determine the relative number of equivalent H atoms represented by each signal.

3. Construct possible molecules from the functional groups present, their relative signal integrations, and any other information you are given, especially the molecular formula and other spectra.

4. Evaluate signal splitting to see if your proposed molecule is consistent with which functional groups are adjacent to each other as evidenced by the splitting.

D. Exceptions to the n + 1 Rule

The general $n + 1$ rule is useful for determining the structures of simple molecules, but *there are many exceptions to this oversimplified rule*. Important examples of exceptions are unsymmetrical molecules with restricted bond rotation such as alkenes or molecule with rings, as well as molecules for which stereochemistry is important (Section 13.11). To understand the complex signal splitting that occurs in these exceptions, the origins of spin-spin coupling must be examined. The following sections of this supplement describe in greater detail spin-spin coupling, and how it leads to complex signal splitting in alkenes and cyclic molecules. For the sake of completeness, signal splitting in alkyl groups is also explained in greater detail.

1.2 A Closer Look at Signal Splitting

A. Review of Spin-Spin Coupling

As described in Section 13.9, ^{1}H-NMR signals are split into multiple peaks when molecules contain **nonequivalent hydrogen atoms** that are separated by no more than three bonds. The multiple peaks are the result of **spin-spin coupling** between ^{1}H nuclei, *an interaction in which nuclear spins of adjacent atoms influence each other*.

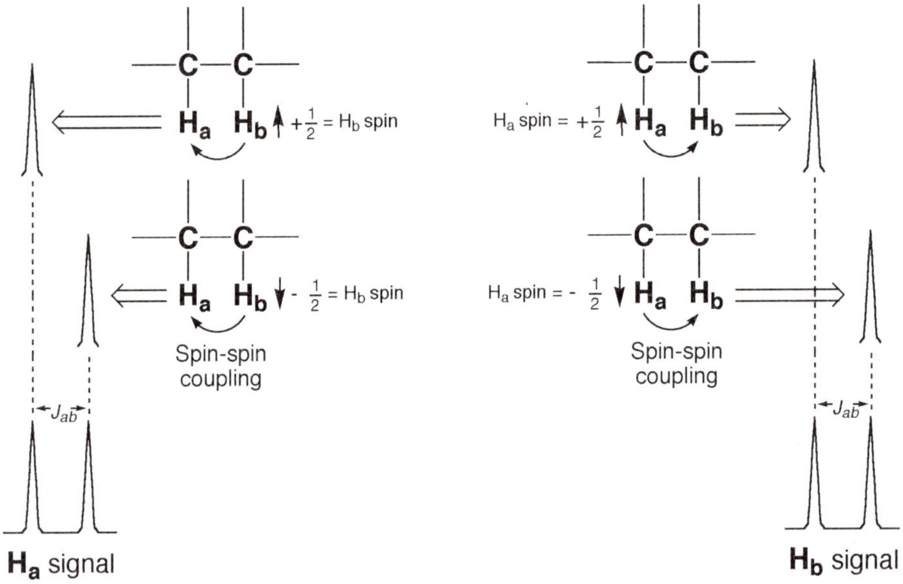

Figure 4 Illustration of spin-spin coupling that gives rise to signal splitting in ^{1}H-NMR spectra. An adjacent nucleus having its spin aligned with an external magnetic field will cause a slightly different signal than an adjacent nucleus having its spin aligned against an external magnetic field. The result is signal splitting, in this case generation of a doublet. Nuclei that are coupled to each other are split with the same coupling constant, referred to as J_{ab} in the figure. In the general case, spin-spin coupling of n adjacent nuclei will cause splitting into $n + 1$ peaks.

The nuclear spin and hence the chemical shift of the atom labeled H_a in Figure 4 is influenced by the adjacent atom H_b, whose nuclear spin might be aligned with or against an external magnetic field in a ^{1}H-NMR spectrometer. Because of spin-spin coupling, alignment of the H_b nuclear spin *with* the external magnetic field results in a slightly different chemical shift of the signal for H_a compared to the situation in which the H_b nuclear spin is aligned *against* the external magnetic field. Across the population of molecules in a sample, there will be similar numbers of molecules having each spin alignment for H_b. Any single molecule gives rise to a single signal for H_a, but the spectrum of the entire sample shows both. The result is that the signal for the H_a atom appears in the spectrum as a **doublet**. In this hypothetical example, the signal for H_b will also be split into a similar doublet due to H_a because the effect operates in both directions. Recall that the **coupling constant, *J*,** is the quantitative measure of the extent of spin-spin coupling between two nuclei and corresponds to the distance in Hz between peaks of a split signal in a ^{1}H-NMR spectrum.

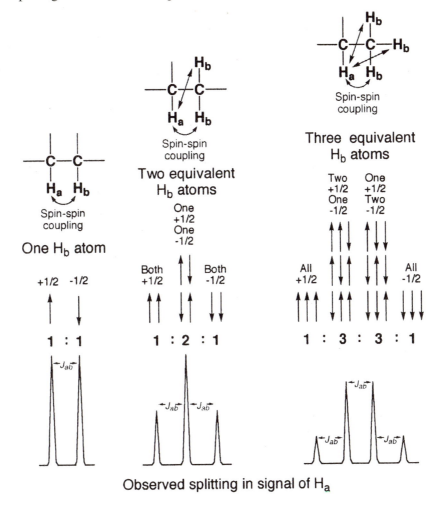

Figure 5 The origins of signal splitting patterns. Each arrow represents an H_b nuclear spin orientation. If there is just one H_b nucleus to consider, there are only two possibilities (↑ or ↓), both of roughly equal probability, leading to a doublet with a 1:1 ratio of peaks for the signal of H_a (left). Two equivalent H_b nuclei can have three different possible combinations that occur in a 1:2:1 ratio (middle), while three equivalent H_b nuclei can have four possible combinations that occur in a 1:3:3:1 ratio (right).

We have already mentioned that in the general case, *n* equivalent H atoms will cause signal splitting into *n* + 1 peaks, the relative intensities of which are predicted by Pascal's triangle. This splitting arises because there are *n* + 1 different possible spin state combinations of *n* spins aligning with or against an external magnetic field. *The probability of a molecule having a given set of spins is proportional to the number of possible spin alignments giving rise to that spin state.* The arrows in Figure 5 are particularly helpful in understanding this very important concept, each arrow representing the spin alignment of a ^{1}H nucleus.

B. Physical Basis

Coupling of nuclear spins is mediated through intervening bonds. The extent of coupling is related to a number of factors, including the number of bonds between the H atoms in question. H atoms with more than three bonds between them generally do not exhibit noticeable coupling, although longer range coupling can be seen in some cases. A common type of coupling involves the H atoms on two C atoms that are bonded to each other. These H atoms are three bonds apart and this type of coupling is referred to as **vicinal coupling**.

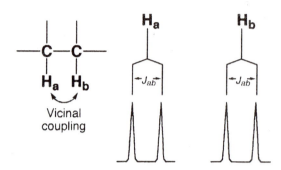

Figure 6 Vicinal coupling between two non-equivalent H atoms.

An important factor in vicinal coupling is the angle α between the C-H sigma bonds and whether or not it is fixed. Coupling is maximized when the angle α is 0° and 180°, and is minimized when α is 90°. Bonds that rotate rapidly at room temperature do not have a fixed angle between adjacent C-H bonds, so an average angle and an average coupling is observed. This latter concept is important for the interpretation of ^{1}H-NMR spectra for alkanes and other flexible molecules.

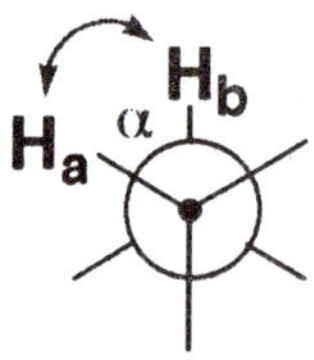

Figure 7 Illustration of the angle α between C-H bonds that is important for determining the strength of spin-spin coupling between adjacent H atoms.

C. More Complex Splitting

So far, we have concentrated on spin-spin coupling with only one other non-equivalent set of H atoms. However, more complex situations often arise in which a set of H atoms is coupled to more than one set of non-equivalent H atoms in molecules that do not have rapid bond rotation. In these situations, the coupling from adjacent non-equivalent sets of H atoms *combine* to give more complex signal splitting patterns. Use of a **tree diagram** is helpful in understanding splitting in these cases. In a tree diagram, the different couplings are applied sequentially.

For example, the atom labeled H_b in Figure 8 is adjacent to non-equivalent atoms H_a and H_c on either side, so the resulting coupling will give rise to a so-called **doublet of doublets**, in other words, a signal with four peaks. Here, the signal for H_b is split into a doublet with coupling constant J_{ab} due to H_a, and this doublet is split into a doublet of doublets with coupling constant J_{bc} by H_c. If there were no other H atoms in the molecule to be considered, then the signal for H_a would be a doublet with coupling constant J_{ab} and the signal for H_c would be a doublet with coupling constant J_{bc}. This analysis assumes that the H_a-H_b and H_b-H_c coupling constants, J_{ab} and J_{bc}, are different from each other. If J_{ab} and J_{bc} are equal, the peaks overlap, a situation discussed in detail later.

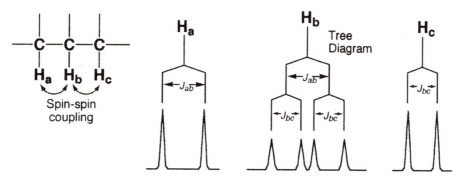

Figure 8 Coupling that arises when H atom H_b is split by two different non-equivalent H atoms H_a and H_c. The spin-spin coupling from H_a and H_c combine to give a doublet of doublets, or four peaks, for the H_b signal. In this example, the signals for H_a and H_c would each be doublets with coupling constants J_{ab} and J_{bc}, respectively. This analysis assumes no other coupling in the molecule and that $J_{ab} \neq J_{bc}$.

If H_c is a set of two equivalent H atoms and H_a is still a single H atom, then the observed coupling would be a doublet of triplets, in other words, a signal with six peaks. Again we are assuming that $J_{ab} \neq J_{bc}$. The tree diagram in Figure 9 shows the complex pattern that results from this type of splitting. If there were no other H atoms in the molecule to be considered, then the signal for H_a would be a doublet with coupling constant J_{ab} and the signal for H_c would be a doublet with coupling constant J_{bc}. The peaks for H_a and H_b would each integrate to a relative value of one H atom, while the peaks for H_c would integrate to a relative value of 2 H atoms.

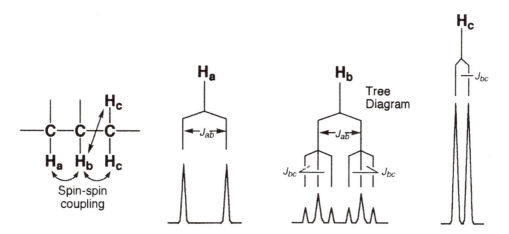

Figure 9 Complex coupling that arises when H atom H_b is split by H_a and two equivalent atoms H_c. The spin-spin coupling from H_a and H_c combine to give a doublet of triplets, or six peaks, for the H_b signal. The signals for H_a and H_c would each be doublets with coupling constants J_{ab} and J_{bc}, respectively. The signal for H_c would integrate to twice the value of H_a or H_b. Again, this analysis assumes no other coupling in the molecule and that $J_{ab} \neq J_{bc}$.

In the general case, a signal will be split into **(n + 1) x (m + 1)** peaks for an H atom that is coupled to a set of **n** H atoms with one coupling constant, and to a set of **m** H atoms with another coupling constant.

D. Bond Rotation

Because the angle between C-H bonds determines the extent of coupling in a molecule, bond rotation is a key parameter. In alkanes and other molecules with free rotation about C-C sigma bonds, H atoms bonded to the same C atom in -CH$_2$- and -CH$_3$ groups are generally equivalent because of the rapid bond rotation. An exception is when a –CH$_2$- is adjacent to a stereocenter, a situation that is discussed in Section 13.11 of the text. However, when there is restricted bond rotation, as in alkenes or cyclic structures, H atoms bonded to the same C atom may not be equivalent, especially if the molecule is not symmetrical. Non-equivalent ^{1}H nuclei on the *same* C atom will couple to each other and cause splitting. This is referred to as **geminal coupling**.

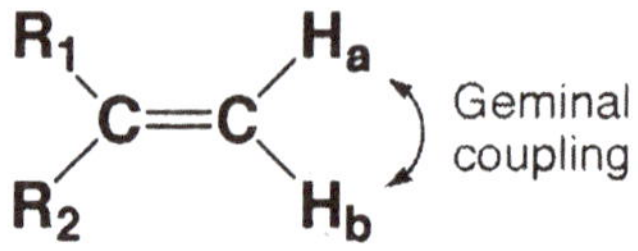

Figure 10 Geminal coupling that occurs when two H atoms on the same carbon atom are not equivalent. This is most common in unsymmetrical alkenes and cyclic molecules.

E. Coincidental overlap

A word of caution. Quite often, because peaks can overlap by coincidence, there are fewer *distinguishable* peaks in a signal than predicted. Coincidental peak overlap can occur in any molecule, but is especially common with flexible alkyl chains. In addition, some coupling constants are so small, that peak splitting is hard to see in a spectrum. Thus, the predicted number of peaks using the (n + 1) x (m + 1) rule should be considered the maximum that *might* be observed. Detailed analysis using extremely high resolution spectrometers is often required to distinguish all of the peaks in a highly split signal.

1.3 Examples of Complex Coupling

A. Alkenes

Characteristic complex coupling can occur in alkenes. Because of the restricted rotation about C=C bonds, the alkenyl (vinylic) H atoms of unsymmetrical alkenes are not equivalent (i.e. they are in unique chemical environments). For example, ethyl propenoate (ethyl acrylate) is an unsymmetrical terminal alkene, so all of the alkenyl H atoms are non-equivalent. As a result, they all couple with each other. In alkenes, *trans* coupling generally results in larger coupling constants (J_{trans} = 11-18 Hz) compared to *cis* coupling (J_{cis} = 5-10 Hz), with geminal coupling being by far the smallest (J_{gem} = 0-5 Hz). Unless a high resolution spectrum is taken, the geminal coupling constant is so small that it is often difficult to see in terminal alkenes. In the spectrum for ethyl propenoate (Figure 11), the geminal coupling is only seen upon close inspection of the signals labeled "a" and "c." You should be able to recognize the characteristic ethyl group pattern of a quartet integrating to two H atoms (-CH$_2$-, H$_d$) and a triplet integrating to three H atoms (-CH$_3$, H$_e$). Tree diagrams are provided in Figure 12 to help decipher patterns of the alkenyl signals.

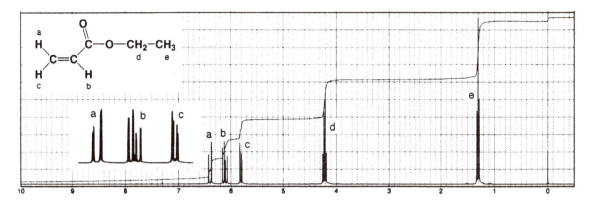

Figure 11 100 MHz ^{1}H-NMR spectrum of ethyl propenoate.

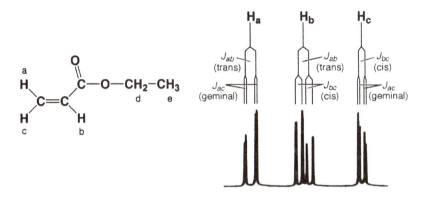

Figure 12 Tree diagrams for the complex coupling seen for the alkenyl H atoms in the ^{1}H-NMR spectrum of ethyl propenoate. All three alkenyl H atoms are non-equivalent and couple with each other. Notice that the *trans* coupling constant, J_{ab}, is the largest, followed by the *cis* coupling constant, J_{bc}. The geminal coupling constant, J_{ac}, is so small that it can only be seen by close inspection of the signals for H_a and H_c.

B. Cyclic structures

Cyclic structures often exhibit restricted rotation about their C-C sigma bonds, and can have constrained conformations. The result is that the two H atoms on –CH$_2$- groups in cyclic molecules can be non-equivalent, leading to complex spin-spin coupling. Substituted epoxides such as 1-methyl-1-vinyloxirane provide a good example (Figure 13). The two H atoms on the three-membered epoxide ring are non-equivalent. H_d is *cis* to the vinyl group and *trans* to the methyl group, while H_e is the reverse. Since they are in different chemical environments, they are non-equivalent and exhibit geminal coupling. The geminal coupling constant is small but discernable in the spectrum as the signals for both H_d and H_e are doublets. Vinyl H atom H_a is split by both H_b (*trans* coupling) and H_c (*cis* coupling), giving rise to a doublet of doublets, or four peaks. H_b is split by H_a (*trans* coupling) along with H_c, the latter geminal coupling constant is so small that it is barely discernable at this resolution. H_c is split by H_a (*cis* coupling) as well as H_b (geminal coupling). The singlet near 1.5 ppm that integrates to three H atoms is the methyl group labeled as "f".

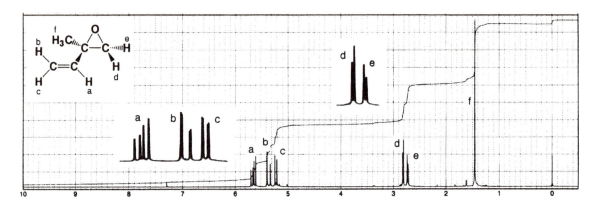

Figure 13 100 MHz ^{1}H-NMR spectrum of 1-methyl-1-vinyloxirane. The two H atoms on the oxirane ring are non-equivalent, so they exhibit geminal coupling.

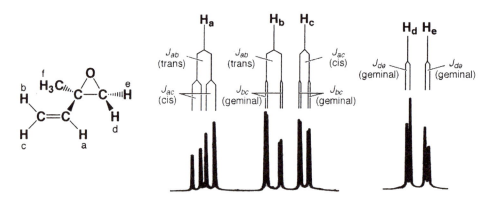

Figure 14 Tree diagrams that indicate the complex coupling seen in ^{1}H-NMR signals for the vinyl group and the oxirane ring H atoms of 1-methy-1-vinyloxirane.

C. Complex Coupling in Flexible Molecules

Coupling in molecules having unrestricted bond rotation is often simplified to the general $n+1$ rule described in Section 1.1B. This can be confusing, because based on the way multiple couplings combine, non-equivalent sets of adjacent H atoms should give rise to more complex couplings than the general $n+1$ rule would predict. The explanation is that bond rotation averages the coupling constants throughout molecules with freely rotating bonds, and tends to make them very similar, in the 6-8 Hz range for H atoms on freely rotating sp^3 hybridized C atoms.

Very similar or identical coupling constants simplify splitting patterns. For example, in the hypothetical unsymmetrical molecule depicted in Figure 15, the central H_b atoms are coupled to both H_a atoms as well as both H_c atoms. If $J_{ab} \neq J_{bc}$, this would lead to a triplet of triplets or nine peaks in the signal for H_b. However, if the coupling constants are identical so that $J_{ab} = J_{bc}$, the splitting pattern overlaps considerably to generate only five peaks in the signal for H_b. In the general case, simplification because of very similar or identical J values gives the familiar $n+1$ peaks when split by combinations of n H atoms, *no matter how many different sets of non-equivalent H atoms are involved.*

A good example is 1-chloro-3-iodopropane. The signal for the H atoms of the central $-CH_2-$ group (labeled as "c" on the spectrum) is split by the H atoms on both of the other $-CH_2-$ groups, raising the possibility of splitting into 3 x 3 = 9 peaks. However, because the values for J_{ab} and J_{bc} are so similar, only 4 + 1 = 5 peaks are distinguishable in the spectrum for the H_c signal due to peak overlap.

12

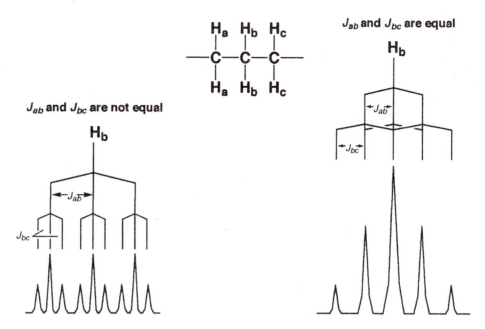

Figure 15 Simplification of signal splitting that occurs when coupling constants are the same. Coupling constants are often similar in alkanes due to rapid bond rotation. In the example, if $J_{ab} \neq J_{bc}$, then a triplet of triplets would be seen in the signal for H_b (left). If $J_{ab} = J_{bc}$, then there is considerable overlap and only 5 peaks are observed (right). When coupling constants are equal, splitting results in $n + 1$ peaks, no matter how many different sets of non-equivalent H atoms are involved.

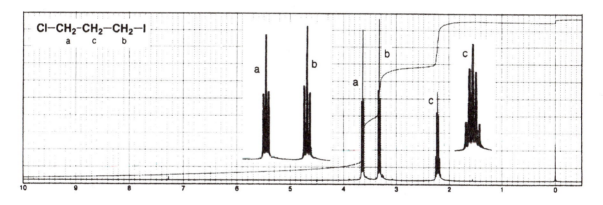

Figure 16 100 MHz ^{1}H-NMR spectrum of 3-chloro-1-iodopropane.

Another common example you have already seen is the splitting of the signal for the central $-CH_2-$ in a $-CH_2-CH_2-CH_3$ group, such as occurs in the molecule 1-chloropropane, $Cl-CH_2-CH_2-CH_3$ (Figure 1). A maximum of $3 \times 4 = 12$ peaks would be possible for the central $-CH_2-$ signal (labeled as "b" in Figure 1), but because the coupling constants are very similar, only $5 + 1 = 6$ peaks are distinguishable.

1.4 Additional Considerations

A. Fast Exchange

Hydrogen atoms bonded to oxygen or nitrogen atoms can exchange with each other faster than the time it takes to acquire a ^{1}H NMR spectrum. This process is greatly facilitated by even traces of acid or base in a sample. Important functional groups affected include carboxylic acids, alcohols, amines, and amides. Fast exchange has two important consequences. First, signals for exchanging H atoms are generally broad singlets that do not take part in splitting with other signals. Second, the signal will disappear altogether if D_2O or a deuterated alcohol is added to the sample because the H atoms will be replaced with D atoms, which are ^1H NMR silent. This latter phenomenon can be used to identify signals from exchangeable H atoms by taking spectra with and without added D_2O. Note that these same exchangeable H atoms also can take part in hydrogen bonds, the presence of which can alter chemical shift in a concentration-dependent fashion.

B. Second-Order Spectra

So far we have only discussed what are generally referred to as **first-order spectra**. Although a detailed discussion is well beyond the scope of this text, it is important to mention briefly that when non-equivalent H atoms are coupled to each other, *and happen to have very similar chemical shifts*, unusual splitting patterns can be observed. This type of peak splitting can range from fewer to more peaks than expected as well as unusual relative peak intensities. Spectra with this unusual splitting are referred to as **second-order spectra**.

C. A Final Word

We have barely scratched the surface of what NMR can do. Something called the nuclear Overhauser enhancement (NOE) can determine distances between atoms in molecules that are near each other in three-dimensional space even if there are more than three bonds separating them. In addition, using modern instruments, spectra can be produced that examine multiple parameters simultaneously to yield immense amounts of information about even very complicated molecules. Such spectra are plotted on more than one axis so they are referred to as multi-dimensional spectra. Multi-dimensional spectra are used to deduce structure and conformation of molecules ranging from small synthetic molecules to large biological macromolecules such as proteins, DNA, and RNA.

1.5 Examples

Following are three ^1H-NMR spectra along with their signal assignments. The spectra are of three carboxylic acids, and have been chosen to illustrate the key issues discussed in this supplement.

A. Pentanoic acid (valeric acid)

Signal "a" – Broad singlet corresponding to the carboxyl H atom near 12 ppm.
Signal "b" – Triplet corresponding to the first $-CH_2-$ group, split by the two "c" H atoms. This is the most downfield $-CH_2-$ signal because these H atoms have the least electron density around them due to being the closest to the electron withdrawing carboxyl group.

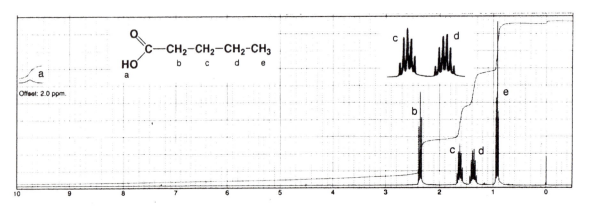

Figure 17 100 MHz ^1H-NMR spectrum of pentanoic acid.

Signal "c" — Quintuplet corresponding to the second –CH$_2$– group, split by the two "b" H atoms as well as the two "d" H atoms. (2 + 2) + 1 = 5 peaks (quintuplet), because $J_{bc} = J_{cd}$.

Signal "d" — Sextuplet corresponding to the third –CH$_2$– group, split by the two "c" H atoms as well as the three "e" H atoms. (2 + 3) + 1 = 6 peaks (sextuplet), because $J_{cd} = J_{de}$.

Signal "e" — Triplet corresponding to the –CH$_3$ group, split by the two "d" H atoms.

B. Trans-2-hexenoic acid

Signal "a" — Broad singlet corresponding to the carboxyl H atom near 12 ppm.

Signal "b" — Doublet of triplets corresponding to alkenyl H atom, split by the *trans* "c" H atom as well as the two "d" H atoms. (1 + 1) x (2 + 1) = 6 peaks (doublet of triplets), since $J_{bc} \neq J_{bd}$.

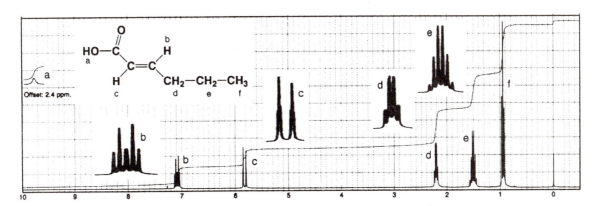

Figure 18 100 MHz ^{1}H-NMR *trans*-2-hexenoic acid.

Signal "c" — Doublet corresponding to the other alkenyl H atom, split by the *trans* "b" H atom with very small triplet splitting from the distant "d" H atoms.

Signal "d" — Four apparent peaks (really a doublet of triplets that overlaps) corresponding to the first –CH$_2$– group, split by the "b" H atom as well as the two "e" H atoms. (1 + 1) x (2 + 1) = 6 peaks (overlapping doublet of triplets), since $J_{bd} \neq J_{de}$. This group is also spit by the distant "c" H atom with a small splitting.

Signal "e" — Sextuplet corresponding to the second –CH$_2$– group, split by the two "d" H atoms as well as the three "f" H atoms. (2 + 3) + 1 = 6 peaks (sextuplet), since $J_{de} = J_{ef}$.

Signal "f" — Triplet corresponding to the –CH$_3$ group, split by the two "e" H atoms.

C. Cyclobutanecarboxylic acid

Here is an example of just how complex splitting can become when geminal H atoms are not equivalent due to restricted bond rotation in a cyclic molecule.

Signal "a" — Broad singlet corresponding to the carboxyl H atom near 12 ppm.

Signal "b" — Quintuplet corresponding to the ring H atom, split by the four "c" H atoms on the ring. The splitting of "b" is likely more complex than this relatively low resolution spectrum indicates, because, in reality, the four different "c" H atoms are not equivalent since two are *cis* to the carboxyl group, and two are *trans* to the carboxyl group on the ring. The signal is relatively far downfield because this H atom is the closest to the electron withdrawing carboxyl group.

Signal "c" — Very complex multiplet corresponding to the four H atoms adjacent to the carbon to which the carboxyl group is bonded. As stated above, these four different "c" H atoms are not equivalent with each other since two are *cis* to the carboxyl group, and two are *trans* to the carboxyl group on the ring. This non-

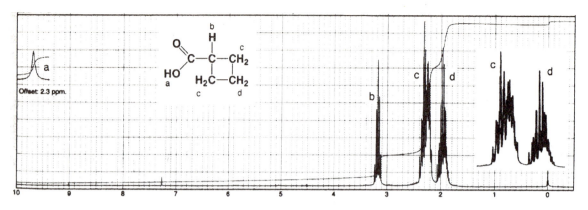

Figure 19 100 MHz ¹H-NMR spectrum of cyclobutanecarboxylic acid.

equivalence gives rise to geminal coupling between the "c" H atoms in addition to vicinal coupling to the "d" H atoms (which are themselves non-equivalent), leading to very complex signal splitting.

Signal "d" — Very complex multiplet corresponding to the two H atoms across the ring from the carboxyl group. These two different "d" H atoms are not equivalent, again because one is *cis* to the carboxyl group, and the other is *trans* to the carboxyl group on the ring. This non-equivalence gives rise to geminal coupling along with vicinal coupling to the "c" H atoms, leading to very complex signal splitting.

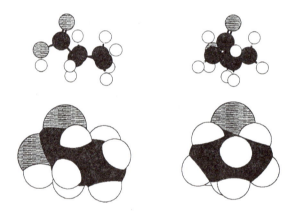

Figure 20 Two views of cyclobutanecarboxylic acid showing how geminal H atoms on the ring are not equivalent to each other. Those on the top face are *cis* to the carboxyl group, and those on the bottom face are *trans* to it. The result is extremely complex splitting patterns in the ¹H-NMR spectrum of Figure 19 as the non-equivalent H atoms split each other through combinations of geminal and vicinal coupling.

On the following pages are MCAT questions that correspond with specific chapters from Brown and Foote's, *Organic Chemistry*, 3e.

Study the chapter then test your knowledge with these sample MCAT questions. Answers and explanations for each question appear on page 72.

How to Prepare for the MCAT

The best preparation for the MCAT starts months before the exam and follows a rigorous course of review and study. For those who wish to begin by seeing what the exam is like and perhaps trying a practice test, this book provides the four full-length practice exams complete with explanatory answers. Visit www.petersons.com bookstore to find more Arco and Peterson's MCAT review books.

Use the scoring guide on the next page to give yourself some indication as to how well you have done on the practice exams. The guide should help you to identify your areas of weakness so that you can direct your study where it will do you the most good. You might also want to highlight your score on each successive practice exam to keep track of your increasing expertise on the MCAT. Do be aware, however, that this guide is only an approximation. This is NOT the actual scoring mechanism used on the MCAT. The range of raw scores contributing to each scaled score is adjusted with each administration of the MCAT to take into account variations in difficulty of questions.

Verbal Reasoning		Physical Sciences		Biological Sciences	
Raw Score	Scaled Score	Raw Score	Scaled Score	Raw Score	Scaled Score
0–6	1	0–9	1	0–8	1
7–14	2	10–17	2	9–17	2
15–21	3	18–28	3	18–29	3
22–28	4	29–34	4	30–34	4
29–34	5	35–39	5	35–40	5
35–38	6	40–45	6	41–46	6
39–43	7	46–49	7	47–51	7
44–48	8	50–56	8	52–58	8
49–53	9	57–61	9	59–64	9
54–56	10	62–65	10	65–68	10
57–59	11	66–69	11	69–70	11
60–61	12	70–72	12	71–72	12
62–63	13	73–74	13	73–74	13
64	14	75–76	14	75–76	14
65	15	77	15	77	15

Periodic Table of the Elements

1 H 1.008																		2 He 4.003
3 Li 6.941	4 Be 9.012											5 B 10.81	6 C 12.01	7 N 14.01	8 O 16.00	9 F 19.00	10 Ne 20.18
11 Na 22.99	12 Mg 24.31											13 Al 26.98	14 Si 28.09	15 P 30.97	16 S 32.06	17 Cl 35.45	18 Ar 39.95
19 K 39.10	20 Ca 40.08	21 Sc 44.96	22 Ti 47.90	23 V 50.94	24 Cr 52.00	25 Mn 54.94	26 Fe 55.85	27 Co 58.93	28 Ni 58.70	29 Cu 63.55	30 Zn 65.38	31 Ga 69.72	32 Ge 72.59	33 As 74.92	34 Se 78.96	35 Br 79.90	36 Kr 83.80
37 Rb 85.47	38 Sr 87.62	39 Y 88.91	40 Zr 91.22	41 Nb 92.91	42 Mo 95.94	43 Tc (98)	44 Ru 101.1	45 Rh 102.9	46 Pd 106.4	47 Ag 107.9	48 Cd 112.4	49 In 114.8	50 Sn 118.7	51 Sb 121.8	52 Te 127.6	53 I 126.9	54 Xe 131.3
55 Cs 132.9	56 Ba 137.3	57 La* 138.9	72 Hf 178.5	73 Ta 180.9	74 W 183.9	75 Re 186.2	76 Os 190.2	77 Ir 192.2	78 Pt 195.1	79 Au 197.0	80 Hg 200.6	81 Tl 204.4	82 Pb 207.2	83 Bi 209.0	84 Po (209)	85 At (210)	86 Rn (222)
87 Fr (223)	88 Ra 226.0	89 Ac** (227)	104 Unq (261)	105 Unp (262)	106 Unh (263)	107 Uns (262)	108 Uno (265)	109 Une (267)									

Alkaline earth metals → (group 2)
Transition Metals
Halogens → (group 17)
Nonmetals / Metals

*Lanthanides (Rare Earths)

58 Ce 140.1	59 Pr 140.9	60 Nd 144.2	61 Pm (145)	62 Sm 150.4	63 Eu 152.0	64 Gd 157.3	65 Tb 158.9	66 Dy 162.5	67 Ho 164.9	68 Er 167.3	69 Tm 168.9	70 Yb 173.0	71 Lu 175.0

**Actinides (Transuranium)

90 Th 232.0	91 Pa (231)	92 U 238.0	93 Np (237)	94 Pu (244)	95 Am (243)	96 Cm (247)	97 Bk (247)	98 Cf (251)	99 Es (252)	100 Fm (257)	101 Md (258)	102 No (259)	103 Lr (260)

CHAPTER 3
CHIRALITY: THE HANDEDNESS OF MOLECULES (4)

1. The structure of 1,1-dichloroethane is shown below.

    If a beam of plane polarized light is passed through a large collection of 1,1-dichloroethane molecules, it would emerge with the plane of polarization having rotated:

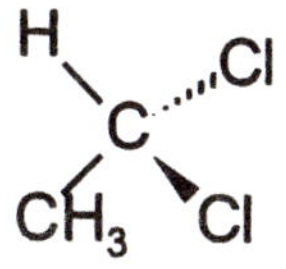

    A. 0°.
    B. 90°.
    C. 60°.
    D. 180°.

2. Consider the following configurations of tartaric acid.

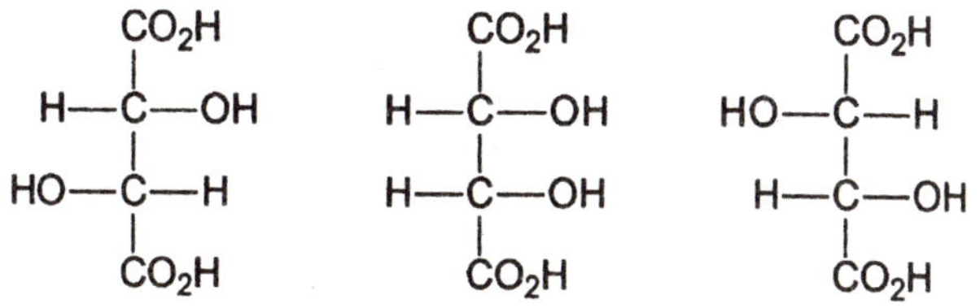

    A diastereomer which has no enantiomer, thus is achiral, is called a *meso* compound. Which of the above configurations would be classified as *meso*?

    A. I only
    B. II only
    C. I and II only
    D. I and III only

3. A compound which has stereocenters yet is achiral, and has no enantiomer, is called a *meso* compound. Which of the following could be classified as a *meso* compound?

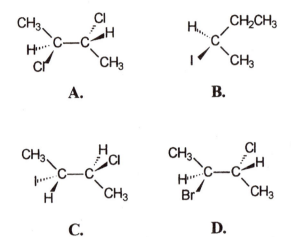

A.   B.   C.   D.

4. Given the structural formula for a compound we can determine:

   I.   the absolute configuration of its stereoisomers.

   II. the direction of optical activity for each stereoisomer.

   III. if it has a meso form.

   A. I and II only
   B. All of the above.
   C. I and III only
   D. II only

CHAPTER 6
ALKENES II (20)

Haloalkenes are produced when hydrogen halides (HX) react with alkenes (I) or by the addition of a halide salt and a strong acid like phosphoric acid (II).

I $CH_3CH=CH_2 + HBr \xrightarrow{HAc} CH_3CHCH_3$
                                              $|$
                                              Br

II cyclohexene $\xrightarrow[H_3PO_4]{KI}$ iodocyclohexane

The mechanism for both I and II involves the protonation of the alkene to form a carbocation followed by the addition of the halide ion.

1. According to Markovnikov's Rule, the electrophilic addition of HBr to 1-butene would yield mostly:

   A. 1-bromobutene.
   B. 2-butene.
   C. 2-bromobutane.
   D. 1,2-dibromobutane.

2. Reactions that follow Markovnikov's Rule are said to proceed with Markovnikov orientation, while those that do not are said to proceed with anti-Markovnikov orientation. The addition of HBr to 3,3,3-trifluoropropene proceeds with anti-Markovnikov orientation. The most likely explanation is:

   A. A carbanion is formed in the rate-determining step.
   B. $H_2\overset{+}{C}-CH_2-CF_3$ is the more stable carbocation.
   C. A bromonium cation adds to the alkene.
   D. The reaction violates mechanistic principles.

3. Hydrogen chloride adds to 3,3-dimethyl-1-butene to give 2-chloro-2,3-dimethylbutane (44% yield) and 2-chloro-2,2-dimethylbutane (37% yield). The best explanation is.

   A. a rearrangement to form a more stable carbocation occurs.
   B. a methide shift occurs.
   C. both A and B are correct.
   D. a different mechanism applies.

4. The addition of hydrogen iodide to 2-butene yields:

   A. a racemic 2-iodobutane solution.
   B. only the R enantiomer.
   C. only the S enantiomer.
   D. 3-butene.

5. Which statement is true concerning the two carbocations formed by the protonation of 2-methylpropene?

   A. The activation energy associated with the primary carbocation intermediate is less than the activation energy associated with the tertiary carbocation.
   B. The rate constant for the formation of the tertiary carbocation is less than the rate constant for the formation of the primary carbocation.
   C. The activation energy and the rate constant associated with the formation of the primary carbocation is greater.
   D. The activation energy is greater for the formation of the primary carbocation but its rate constant is less.

A student found that when she reacted 2-methyl-1-butanol with H₂SO₄, two low-boiling liquids formed (Solution A). She then performed the following experiments.

Experiment 1

The liquids were reacted with Cl₂ in CCl₄ and the resulting mixture distilled. Two dichloro products were identified as mostly 2,3-dichloro-2-methylbutane and some 1,2-dichloro-2-methylbutane.

Experiment 2

The liquids in Solution A were then reacted with HCl(aq) to form mostly 2-chloro-2-methylbutane.

The student concluded that the liquids were 2-methyl-1-butene and 2-methyl-2-butene. She performed Experiments 3 and 4 to elucidate the mechanism of the reaction.

Experiment 3

A kinetic study showed that the initial reaction is a first-order reaction whose rate depends only on the concentration of 2-methyl-1-butanol.

Experiment 4

The initial reaction was repeated using 2-deutero-2-methyl-1-butanol as a starting material. The rate of the reaction decreased as a result.

6. What are the intermediates that could account for the two liquid products?

   A. CH₃—⁺CH—CH(CH₃)—CH₃, ⁺CH₂—CH₂—CH(CH₃)—CH₃

   B. ⁺CH₂—CH₂—CH(CH₃)—CH₃, CH₃—CH₂—⁺C(CH₃)—CH₃

   C. CH₃—CH₂—CH(CH₃)—⁺CH₂, CH₃—CH₂—⁺C(CH₃)—CH₃

   D. CH₃—⁺C(CH₃)(CH₃)—CH₂, CH₃—⁺C(CH₃)—CH₂—CH₃

7. Which mechanism of the initial reaction is most consistent with the results of the experiments?

   A. S$_N$1
   B. S$_N$2
   C. E1
   D. E2

8. Chlorination of the liquids showed that Solution A consisted mostly of

   A. 2-methyl-2-butene.
   B. 2-methyl-1-butene.
   C. 1-chloro-2-methyl-2-butene.
   D. 1-chloro-2-methyl-1-butene.

9. The results of the experiments are consistent with a carbocation undergoing a

   A. 1,2 methyl shift.
   B. 1,2 hydride shift.
   C. 1,3 methyl shift.
   D. 1,3 hydride shift.

10. Based on the results of Experiments 1–4, what is most likely the chief product of the dehydrohalogenation of 1-bromo-2-methylpentane?

    A. 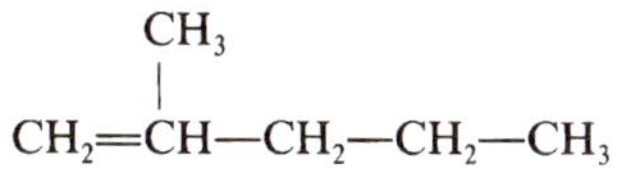 $CH_2{=}C(CH_3){-}CH_2{-}CH_2{-}CH_3$

    B. $CH_3{-}C(CH_3){=}CH{-}CH_2{-}CH_3$

    C. $CH_3{-}CH(CH_3){-}CH{=}CH{-}CH_3$

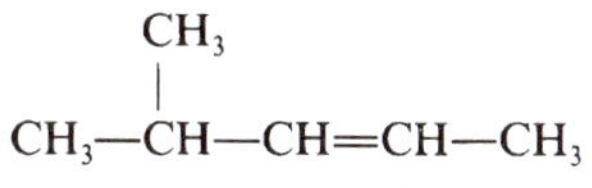

    D. 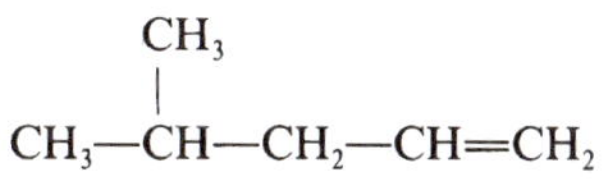 $CH_3{-}CH(CH_3){-}CH_2{-}CH{=}CH_2$

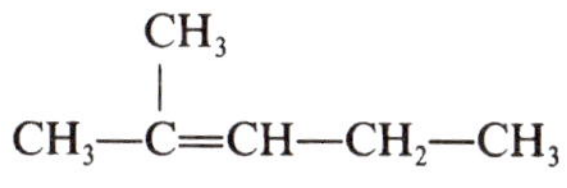

11. Which of the following general statements applies to the mechanism of the reaction?

    A. An S_N2 reaction occurs with inversion of configuration.
    B. Rearrangement of $H^\ominus$ or an alkyl group will occur if it produces a more stable carbocation.
    C. Electron-withdrawing groups stabilize carbocations.
    D. Only cis isomers are formed in elimination reactions.

Passage I (Questions 12–19)

Two mechanisms are postulated for the addition of Br_2 to cis-2-butene.

Mechanism A

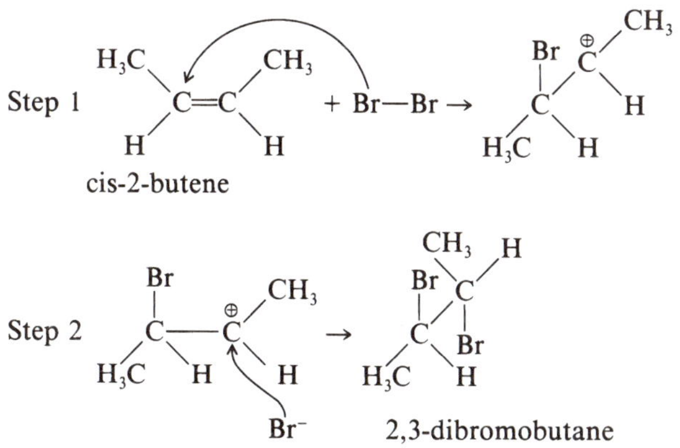

Step 1: cis-2-butene + Br—Br → bromonium-like intermediate

Step 2: intermediate + Br⁻ → 2,3-dibromobutane

25

Mechanism B

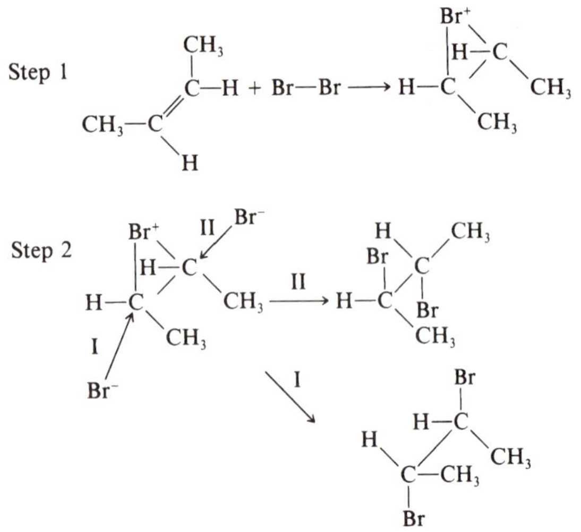

Step 1

Step 2

Bromination of cis-2-butene yields a racemic mixture of 2,3-dibromobutane while bromination of trans-2-butene gives meso-2,3-dibromobutane.

12. Step 2 of Mechanism A as shown yields only

   A. (R,R)-2,3-dibromobutane.
   B. (S,S)-2,3-dibromobutane.
   C. (R,S)-2,3-dibromobutane.
   D. meso-2,3-dibromobutane.

13. Step 2 of Mechanism A shows attack of Br from the bottom of the carbocation.

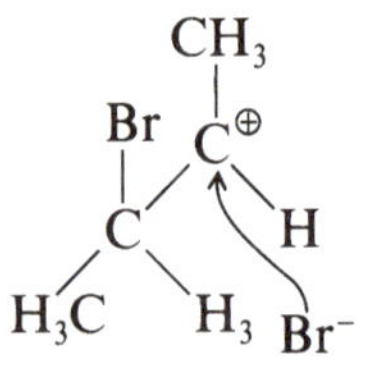

   If the rotation about the CC bond occurs so that the carbocation in Step 2 of Mechanism A becomes

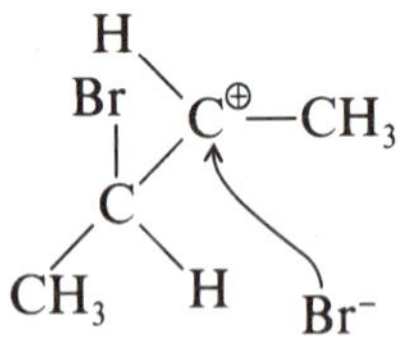

   the product would be

   A. meso-2,3-dibromobutane.
   B. the same product.
   C. (S,S)-2,3-dibromobutane.
   D. (R,R)-2,3-dibromobutane.

14. The observed stereochemistry

   A. supports both Mechanism A and Mechanism B.
   B. does not support either mechanism.
   C. supports Mechanism A only.
   D. supports Mechanism B only.

15. The advantage of a bromonium ion over a carbocation as an intermediate is

   A. that in a carbocation, all atoms have an octet.
   B. that in a bromonium ion, all atoms have an octet.
   C. that a halogen can better accommodate the positive charge.
   D. There is no advantage.

16. Bromination of 2-phenyl-2-butene occurs via Mechanism A. What is the most likely explanation?

   A. Phenyl groups undergo addition more readily.
   B. The bromonium ion is sterically hindered.
   C. The phenyl group stabilizes the carbocation.
   D. The phenyl group destabilizes the bromonium ion.

17. Halogenation of both cis and trans-2-butene occurs by

   A. syn addition.
   B. anti addition.
   C. syn addition for cis-2-butene and anti for trans.
   D. anti addition for cis-2-butene and syn for trans.

18. Halohydrins contain halogen and hydroxyl groups on adjacent carbon atoms. They can be formed by the addition of chlorine or bromine to an alkene in the presence of water. There is evidence that a halonium ion is formed as an intermediate, which is then attacked by H₂O. Predict the major product(s) when cis-2-butene forms a chlorohydrin.

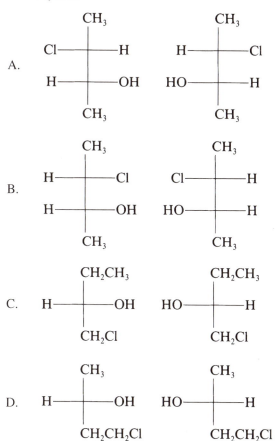

19. Hydroxylation of cis-2-butene with peroxy acids forms a racemic mixture of 2,3-butanediol. Which intermediate is most consistent with the stereochemical evidence?

   A. a carbocation

   B. an epoxide

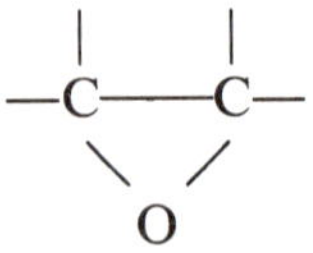

   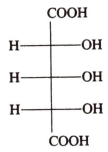

   C. a carbanion
   D. a halonium ion

20. Which of the statements below is false regarding the following structure?

   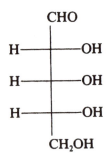

   A. It is optically inactive.
   B. It is a meso acid.
   C. It can be formed by using HNO_3 to treat

   $$\begin{array}{c} CHO \\ H-\!\!\!\!\!-\!\!\!\!\!-OH \\ H-\!\!\!\!\!-\!\!\!\!\!-OH \\ H-\!\!\!\!\!-\!\!\!\!\!-OH \\ CH_2OH \end{array}$$

   D. It is nonsuperimposible on its mirror image.

CHAPTER 8
NUCLEOPHILIC SUBSTITUTION AND β-ELIMINATION (18)

1. The reaction between a secondary alkyl halide and a base would tend to yield a higher percentage of elimination product by treatment with:

    A. a very weak base which is also a very hindered electrophile.
    B. a very weak base which is also a very hindered nucleophile.
    C. a very strong base which is also a very hindered electrophile.
    D. a very strong base which is also a very hindered nucleophile.

Passage I (Questions 2–8)

When a compound reacts with the solvent, the process is referred to as a *solvolysis* reaction. A chemist studied the mechanisms of the solvolysis reactions of t-butyl bromide (t-BuBr) in pure ethanol, and in a mixed solvent system containing 80% ethanol and 20% water.

In Reaction I, t-BuBr was refluxed in ethanol containing an added portion of sodium ethoxide. The reaction was exceedingly rapid with a half life of only a few minutes. The solvolysis product A was produced along with a significant amount of product B. Compound A was found to be an ether that had a molecular formula of $C_6H_{14}O$. Compound B had a molecular formula of C_4H_8 and the ^{1}HNMR revealed the presence of vinylic protons.

The rate of reaction of t-butyl bromide is given by the following expression:

rate = k_1 [t-BuBr] + k_2 [t-BuBr][$C_2H_5O^-$]

Gaseous hydrogen chloride was bubbled slowly into a solution of Compound B in methylene chloride at 0 °C. This reaction yielded t-butyl chloride (Compound D).

In Reaction II, t-BuBr was refluxed in the mixed solvent system of 80% ethanol and 20% water. Three compounds were isolated: A, B and C.

The IR spectrum of Compound C indicated a very intense stretch centered at about 3,600 cm^{-1}. Compound C had a molecular formula of $C_4H_{10}O$. A mixture of Compound C and PBr$_3$ was allowed to react further, resulting in the reformation of t-BuBr.

2. What is the IUPAC name of Compound B?

    A. 3,4-dimethylpropane
    B. 2-methyl-1-propene
    C. 2-methyl-2-butene
    D. propene

3. What functional group is present in Compound C?

    A. Ether
    B. Carboxylic acid
    C. Hydroxyl
    D. Ketone

4. The rates of solvolysis of t-butyl chloride, bromide, and iodide are predicted to be different. Which of the following series best represents the arrangement of these tertiary alkyl halides in order of decreasing rates of solvolysis?

   A. t-BuBr > t-BuI > t-BuCl
   B. t-BuCl > t-BuI > t-BuBr
   C. t-BuCl > t-BuBr > t-BuI
   D. t-BuI > t-BuBr > t-BuCl

5. The solvolysis reaction of t-butyl bromide in pure ethanol to produce Compound A can be classified mechanistically as:

   A. bimolecular nucleophilic substitution.
   B. bimolecular electrophilic substitution.
   C. unimolecular electrophilic substitution.
   D. unimolecular nucleophilic substitution.

6. What type of intermediate was formed in both reactions I and II?

   A. A carboxylic anion
   B. A free radical
   C. A carbocation
   D. A carbanion

7. In Reaction II, t-butyl bromide gave 3 products 3 reaction mechanisms.

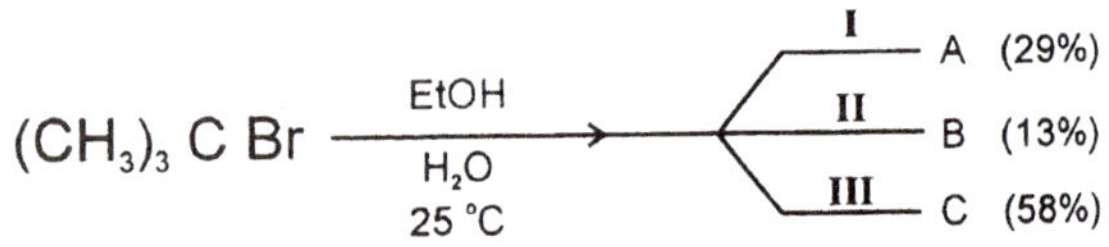

   $(CH_3)_3 C Br \xrightarrow[H_2O]{EtOH, 25°C}$
   I → A (29%)
   II → B (13%)
   III → C (58%)

   Which of the following series is an acceptable mechanistic rationalization for processes I, II and III leading to the formation of compounds A, B, and C respectively?

   A. S_N1, S_N1, E2
   B. S_N1, E2, S_N1
   C. S_N2, S_N1, E1
   D. S_N2, S_N2, S_N1

8. Which of the following is the most accurate representation of the reaction coordinate diagram for the solvolysis of t-butyl bromide?

   A.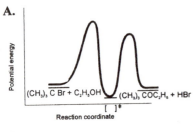
   B.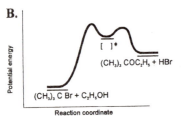
   C.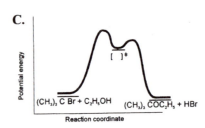
   D.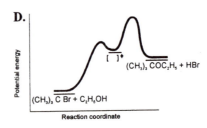

Passage II (Questions 9–12)

A chemist studied the mechanisms of the reactions of isopropyl bromide with sodium t-butoxide and with sodium ethoxide.

In Reaction I, the treatment of isopropyl bromide with sodium t-butoxide at 40 °C gave almost exclusively one product. The reaction yielded Compound A which had a molecular formula of C_3H_6. The ^{1}HNMR spectrum of compound A revealed the presence of vinylic protons. The kinetic rate expression indicated second order kinetics for the reaction.

In Reaction II, the treatment of isopropyl bromide with sodium ethoxide (Na^+EtO^-) at 30 °C yielded an ether (Compound B) that had a molecular formula of $C_5H_{12}O$, and Compound A that had a molecular mass of 42 grams. Compound B accounted for only 20% of the total product. The remainder consisted of Compound A.

Compound B was found to be stable to base, dilute acid, and most reducing agents. The infrared spectrum of Compound B revealed a strong band at 1100 cm^{-1} and the absence of stretching absorptions at 1730 cm^{-1} was noted.

Compound A was readily oxidized by a neutral solution of cold dilute potassium permanganate. During the oxidation process, the characteristic purple color of the permanganate ion (MnO_4^-) disappeared and was replaced by a brown precipitate indicating the formation of manganese dioxide (MnO_2).

9. Which of the following best describes the reaction mechanisms that led to the formation of products A and B?

    A. B is the S_N2 product and A is derived from an E2 reaction.
    B. A is the S_N2 product and B is derived from an E2 reaction.
    C. B is the S_N1 product and A is derived from an E1 reaction.
    D. B is the S_N1 product and A is derived from an E2 reaction.

10. Compound A belongs to which of the following classes of organic compounds?

    A. Alcohol
    B. Ketone
    C. Alkene
    D. Ester

11. Which of the following most accurately represents the activated complex formed in Reaction II and that subsequently led to Compound A?

    A. $H-\underset{CH_3}{\overset{CH_3}{|}}C\cdots Br$

    B. $H-\underset{CH_3}{\overset{CH_3}{|}}C\oplus$

    C. $Et-O\overset{\delta-}{\cdots}H\cdots\underset{H}{\overset{H}{|}}C\cdots\underset{H}{\overset{CH_3}{|}}C\overset{\delta-}{\cdots}Br$

    D. $Et-O\overset{\delta-}{\cdots}\underset{CH_3}{\overset{CH_3}{\underset{|}{|}}}C\overset{\delta-}{\cdots}Br$ (with H)

12. WHich of the following compounds is an accurate representation of Compound B?

    A. CH_3OCH_3
    B. $C_2H_5OC_2H_5$
    C. $(CH_3)_2CHOCH_2CH_3$
    D. $CH_3CH_2CH_2OCH_2CH_3$

Passage III (Questions 13–18)

An organic chemistry student was assigned to review two important preparations of alkenes; the dehydration of alcohols and the dehydrohalogenation of alkyl halides.

Experiment I

   The treatment of isopropyl iodide with potassium ethoxide in ethanol gave propylene in 94% yield. The reaction revealed first order kinetics in alkyl halide and first order kinetics in base.

Experiment II

   1-Bromopentane was treated with potassium hydroxide in ethanol to yield both 1-pentene and ethyl pentyl ether, in 12% and 88% yields, respectively.

Experiment III

   Ethyl alcohol was treated with sulfuric acid at 140 °C to produce diethyl ether.

However, when ethyl alcohol was treated with sulfuric acid at an elevated temperature of 170 °C, ethylene was the major roduct.

13. Which of the following reaction mechanisms best exemplifies the mechanistic process in Experiment I?

    A. E1
    B. E2
    C. S_N1
    D. S_N2

14. In a subsequent experiment the student performed the catalytic hydrogenation of propylene by using hydrogen at low temperature in the presence of a platinum catalyst. Which of the following would likely be the major product?

    A. 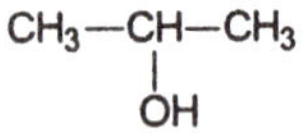 $CH_3-\underset{\underset{OH}{|}}{CH}-CH_3$

    B. 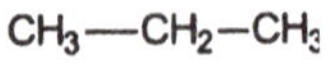 $CH_3-CH_2-CH_3$

    C. 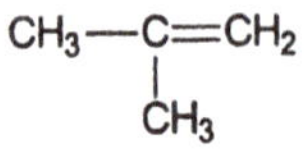 $CH_3-\underset{\underset{CH_3}{|}}{C}=CH_2$

    D. 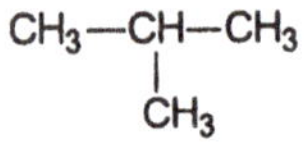 $CH_3-\underset{\underset{CH_3}{|}}{CH}-CH_3$

32

15. Treatment of ethylene with cold dilute potassium permanganate would yield which of the following compounds?

A. 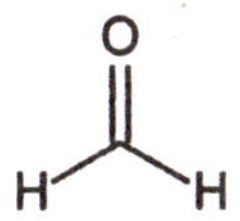 H–CHO (formaldehyde)
B. 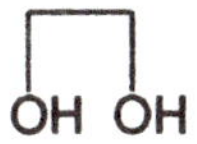 HOCH₂CH₂OH (structure with two OH groups on adjacent carbons)
C. cyclic structure with O–O
D.  epoxide (oxirane)

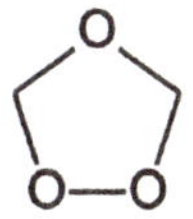

16. In Experiment II, ethyl pentyl ether was the major product. Which of the following assumptions can be made about the mechanism of this reaction?

   A. The reaction proceeded by nucleophilic substitution via the S_N2 mechanism.
   B. The reaction proceeded by nucleophilic substitution via the S_N1 mechanism.
   C. The reaction proceeded by electrophilic substitution via the E1 mechanism.
   D. The reaction proceeded by electrophilic substitution via the E2 mechanism.

17. Consider the following reaction:

   $(CH_3)_2C = CH_2 + HCl \rightarrow$ Product

   Which of the following compounds best exemplifies the major organic product of the above reaction?

   A. $(CH_3)_2 CHCH_2 Cl$

   B. 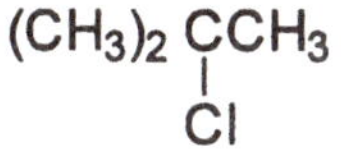 $(CH_3)_2 C(Cl)CH_3$

   C. $CH_2=C(Cl)-CH_2CH_3$

   D. 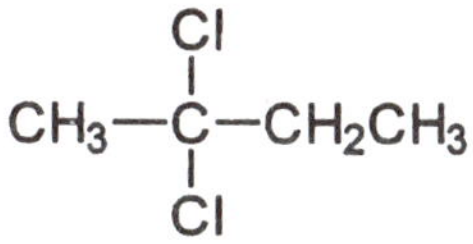 $CH_3-C(Cl)_2-CH_2CH_3$

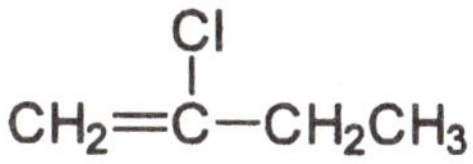

18. An organic chemist investigated the mode of addition of HBr to terminal olefins via two different mechanisms.

   $CH_3 CH_2 CH = CH_2 + HBr \xrightarrow{\text{ionic mechanism}}$ A   Reaction I

   $CH_3 CH_2 CH = CH_2 + HBr \xrightarrow{\text{free radical mechanism}}$ B   Reaction II

   Which of the following pairs of compounds represents major products A and B, respectively?

   A. 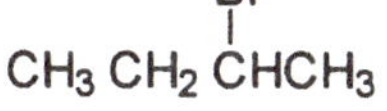 $CH_3 CH_2 CHBrCH_3$, $CH_3 CH_2 CHBrCH_3$

   B. $CH_3 CH_2 CH_2 CH_2 Br$, $BrCH_2 CH_2 CH_2 CH_3$

   C. $BrCH_2 CH_2 CH_2 Br$, $CH_3 CH_2 CHBrCH_3$

   D. 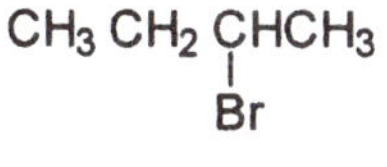 $CH_3 CH_2 CHBrCH_3$, $CH_3 CH_2 CH_2 CH_2 Br$

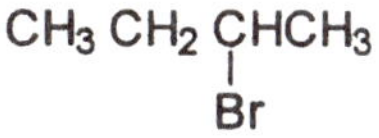

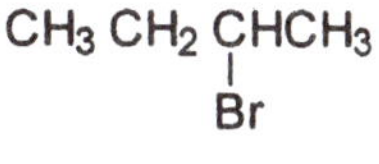

33

CHAPTER 9
ALCOHOLS AND THIOLS (2)

1. Acid catalysts such as *p*-toluenesulfonic acid are often used to dehydrate alcohols. The role of the acid catalyst is to:

    A. increase $\Delta G°$ and increase the activation energy for the dehydration reaction.
    B. increase $\Delta G°$ and lower the activation energy for the dehydration reaction.
    C. maintain $\Delta G°$ at the same value and lower the activation energy for the dehydration reaction.
    D. lower $\Delta G°$ and increase the activation energy for the dehydration reaction.

2. Chromatography can be used to:

    A. separate nonvolatile liquids.
    B. separate volatile liquids.
    C. separate a nonvolatile liquid from a volatile liquid.
    D. All of the above

CHAPTER 13
NUCLEAR MAGNETIC RESONANCE SPECTROSCOPY

1. Four compounds—allyl alcohol, benzoic acid, 2-butanone and butyraldehyde—were identified and stored in separate bottles. By accident, the labels were lost from the sample bottles. The following information was obtained via infrared spectroscopy and was used to identify and relabel the sample bottles.

   Infrared absorption peaks (cm^{-1})

Bottle I	Bottle II	Bottle III	Bottle IV
1700 (sharp)	1710	1730 (sharp)	3333 (broad)
	3500–3333 (broad)	2730	1030 (small)

   Which of the following most accurately represents the contents of bottles I, II, III, and IV, respectively?

   A. Butyraldehyde, 2-butanone, benzoic acid, allyl alcohol
   B. Benzoic acid, butyraldehyde, allyl alcohol, 2-butanone
   C. 2-Butanone, butyraldehyde, allyl alcohol, benzoic acid
   D. 2-Butanone, benzoic acid, butyraldehyde, allyl alcohol

2. The ^{1}HNMR of compound X is shown in the following figure.

   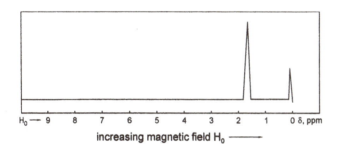

   Compound X has the molecular formula C_6H_{12} and chemical tests revealed that the compound was an alkene. Which of the following best represents the structure of compound X?

   A. $CH_3-CH=CH-CH_2-CH_2-CH_3$

   B. 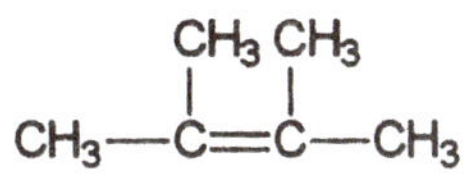 $CH_3-\underset{\underset{CH_3}{|}}{C}=\underset{\underset{CH_3}{|}}{C}-CH_3$

   C.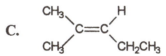

   D. 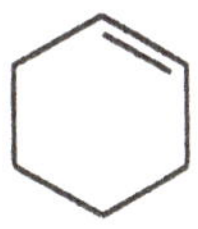 (cyclohexene)

Passage I (Questions 3–7)

An organic chemistry student conducts a free radical reaction by exposing a mixture of 2-methylpropane and Br_2 to light. She separates the products and obtains the NMR spectrum of a dibromo substituted propane. The spectrum contains two singlets with a peak ratio of 3:1.

3. The NMR spectrum is of:

    A. 1,1-dibromo-2-methylpropane.
    B. 1,3-dibromo-2-methylpropane.
    C. 1,2-dibromo-2-methylpropane.
    D. 2-bromo-2-methylpropane.

4. A dehydrohalogenation reaction produces 2-methyl-3-bromo-1-propene. How many signals will its NMR spectrum contain?

    A. 4
    B. 3
    C. 2
    D. 1

5. The NMR spectrum of 1,1-dideutero-2-methyl-3-bromo-1-propene contains:

    A. a quartet and a triplet.
    B. two singlet peaks.
    C. three singlet peaks.
    D. a singlet, a quartet, and a triplet.

6. The effect of bromine on the chemical shift of a proton bonded to the same carbon is to:

    A. move the absorption downfield.
    B. move the absorption upfield.
    C. cause a singlet peak to split into a doublet peak.
    D. increase the intensity of the peak.

7. Which of the following statements is FALSE?

    A. The area under an NMR signal depends on the number of protons causing the signal.
    B. Splitting of NMR signals is caused by nearby protons.
    C. Equivalent protons have the same chemical shift.
    D. Only protons give rise to NMR spectra.

Passage II (Questions 8–12)

The infrared spectrum of an organic compound composed of carbon, oxygen, and hydrogen shows strong absorptions at about 3350 cm^{-1}; 1000 cm^{-1}; and 675–870 cm^{-1}. The absorption at 3350 cm^{-1} is both strong and broad. The compound also exhibits strong absorption bands in the ultraviolet region.

Infrared Absorption by Some Oxygen-containing Compounds

Compound	O—H	C—O	C=O
Alcohols	3200–3600 cm^{-1}	1000–1200 cm^{-1}	—
Phenols	3200–3600	1140–1230	—
Ethers, aliphatic	—	1060–1150	—
Ethers, aromatic	—	1200–1275	—
	1020–1075		
Aldehydes, ketones	—	—	1675–1725 cm^{-1}
Esters	—	1250 (*two bands*)	1680–1725
Acid chlorides	—	—	1750–1810
Amides (RCONH$_2$)	(N—H 3050–3550)	—	1650–1690

8. Based on the table, the infrared spectrum indicates which of the following oxygen-containing functional groups is(are) present?

   A. Carbonyl and hydroxyl
   B. Hydroxyl and ether
   C. Carboxyl group
   D. Hydroxyl group

9. Absorption in the ultraviolet region indicates the presence of:

   A. a conjugated structure or aromatic group.
   B. an alkyl group.
   C. one double bond.
   D. an ether group.

10. The NMR spectrum contains several signals, one of which exhibits splitting only when a dry, pure sample is used. The splitting disappears on the addition of a trace of acid or base. This same signal is absent in the NMR spectrum of the product formed by reaction with a carboxylic acid. The functional group that can account for this behavior is which of the following?

    A. H—C(=O)—
    B. —C(=O)—OH
    C. —C(OH)—
    D. —C—O—C— (with H on one C)

37

11. The infrared spectrum of the compound also contains moderate and weak bands associated with a mono-substituted benzene. Which of the following is the most likely structure of the compound?

    A. 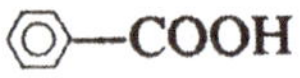 ⌬—COOH
    B.  ⌬—CH₂OH
    C.  ⌬—O—CH₃
    D. 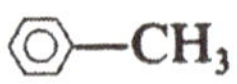 ⌬—CH₃

12. Which of the following spectroscopic techniques should be used to establish the presence of a carbonyl group in an organic molecule?

    A. NMR
    B. Ultraviolet spectroscopy
    C. Infrared spectroscopy
    D. Mass spectrometry

CHAPTER 16
ALDEHYDES AND KETONES (2)

1. When a dilute solution of formaldehyde is dissolved in ^{18}O-labeled water and allowed to equilibrate, ^{18}O incorporation occurs thus indicating the existence of an intermediate product. Which of the following compounds best represents the intermediate product of this ^{18}O exchange?

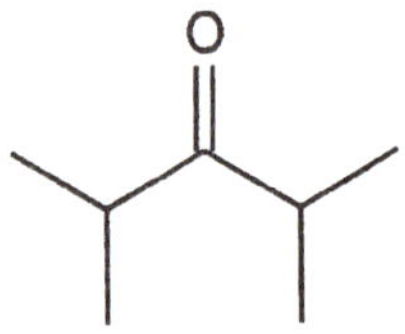

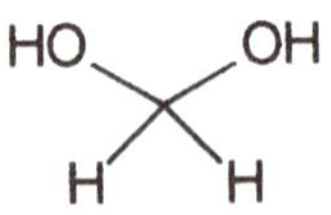

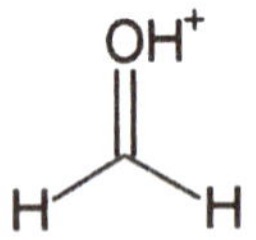

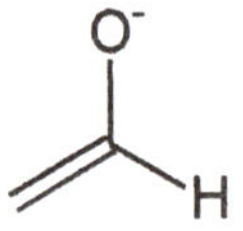

2. Aldehydes react with alcohols in acidic solution to form:

    A. hemiacetals and acetals.
    B. hemiacetals only.
    C. ketals only.
    D. hemiketals only.

CHAPTER 18
FUNCTIONAL DERIVATIVES OF CARBOXYLIC ACIDS

Passage I (Questions 1–6)

These experiments were performed to study esterification reactions.

Experiment 1

Labeled $CH_3^{18}OH$ reacted with acetic acid to form methyl acetate enriched in ^{18}O:

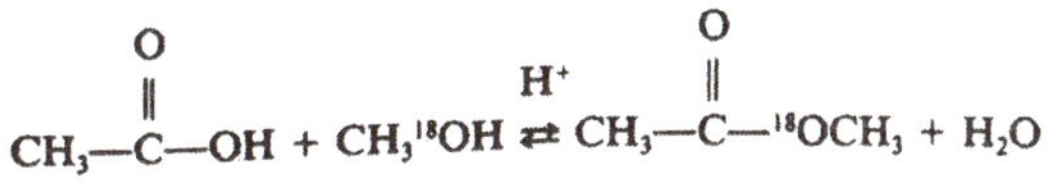

$$CH_3-\underset{\underset{O}{\|}}{C}-OH + CH_3{}^{18}OH \rightleftarrows CH_3-\underset{\underset{O}{\|}}{C}-{}^{18}OCH_3 + H_2O$$

Experiment 2

Acetic acid was reacted with several alcohols. The rate of esterification increased according to

$$(CH_3)_3C-OH < (CH_3)_2CH-OH < CH_3CH_2OH < CH_3OH$$

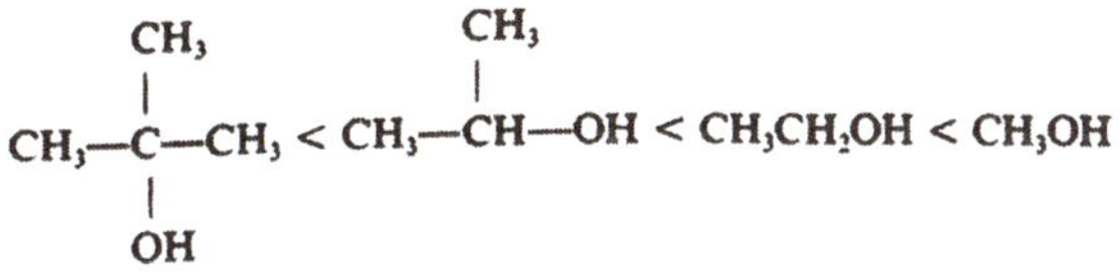

Experiment 3

Ethanol was reacted with several carboxylic acids. The rate of esterification increased according to
$(CH_3)_3CCOOH < (CH_3)_2CHCOOH < CH_3CH_2COOH < CH_3COOH$

1. The results of Experiment 1 indicate that in esterification the bond that is cleaved in the carboxylic acid, $R-\underset{\underset{O}{\|}}{C}-OH$, is which of the following?

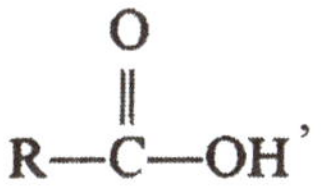

    A. C⫤OH
    B. R⫤C
    C. CO⫤H
    D.  $\underset{C}{\overset{O}{⫤}}$

2. The mechanism for the acid-catalyzed hydrolysis of ester $R-\underset{\underset{O}{\|}}{C}-OR'$, involves the breaking of which of the following bonds?

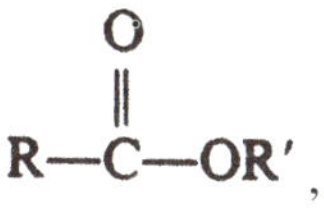

    A. R⫤C
    B. C⫤OR'
    C. CO⫤R'
    D.  $\underset{C}{\overset{O}{⫤}}$

3. Strong acids catalyze both esterification and hydrolysis because:

    A. they add to the carbonyl carbon.
    B. they protonate the alcoholic oxygen.
    C. the carbonyl carbon is made more susceptible to attack by a nucleophile when the carbonyl oxygen is protonated.
    D. they remove OH– ions.

4. The results of Experiments 2 and 3 support which of the following statements?

    A. The presence of bulky groups near the site of the reaction slows both esterification and hydrolysis reactions.
    B. The presence of bulky groups on only the alcohol slows both esterification and hydrolysis reactions.
    C. The presence of bulky groups near the site of the reaction slows esterification with no effect on hydrolysis.
    D. The presence of bulky groups near the site of the reaction slows hydrolysis with no effect on esterification.

5. Basic hydrolysis of esters involves attack of OH$^-$ and is essentially irreversible.

    $$R-\overset{O}{\underset{\|}{C}}-OR' + OH^- \rightarrow R-\overset{O}{\underset{\|}{C}}-O^{\ominus} + R'OH$$

    A basic hydrolysis reaction is carried out using a labeled ester:

    $$CH_3-\overset{O}{\underset{\|}{C}}-{}^{18}OC_2H_5 + OH^- \rightarrow CH_3-\overset{O}{\underset{\|}{C}}-O^{\ominus} + C_2H_5{}^{18}OH$$

    Therefore, in basic hydrolysis of an ester, cleavage occurs between:

    A. oxygen and the alkyl group of the alcohol.
    B. the carbonyl carbon and the alkyl group of the acid.
    C. the carbonyl carbon and the carbonyl oxygen.
    D. oxygen and the acyl group.

6. Esterification and both acidic and basic hydrolysis reactions can be described as attack of:

    A. a nucleophile on the carbonyl carbon.
    B. a nucleophile on an alkyl carbon.
    C. an electrophile on the carbonyl carbon.
    D. an electrophile on an alkyl carbon.

7. Which of the following compounds will not react with acetyl chloride?

    A. NH_3
    B. CH_3NH_2
    C. $(CH_3)_2NH$
    D. $(CH_3)_4N^+Cl^-$

8. Basic hydrolysis of the dipeptide

Ala-Gly : H$_2$N—CH(CH$_3$)—C(=O)—N(H)—CH(H)—COOH gives:

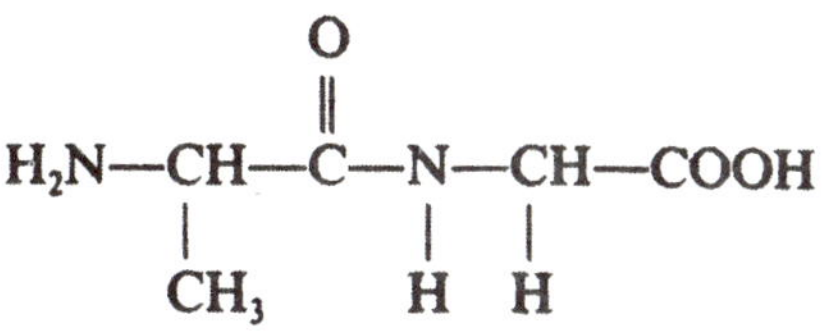

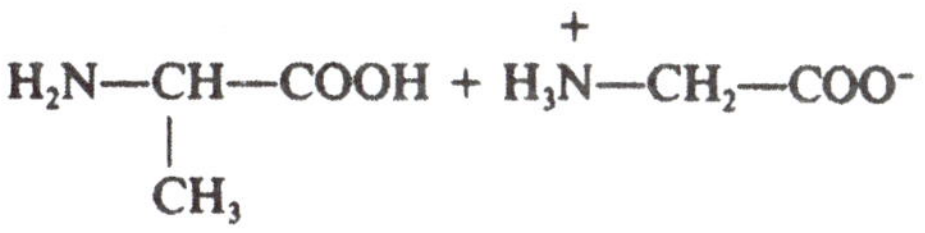

A. 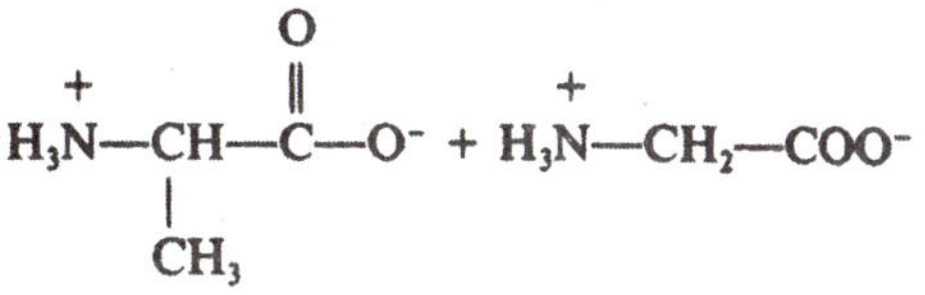 $^+$H$_3$N—CH(CH$_3$)—C(=O)—O$^-$ + $^+$H$_3$N—CH$_2$—COO$^-$

B. H$_2$N—CH(CH$_3$)—COOH + $^+$H$_3$N—CH$_2$—COO$^-$

C. H$_2$N—CH(CH$_3$)—COO$^-$ + H$_2$N—CH$_2$—COO$^-$

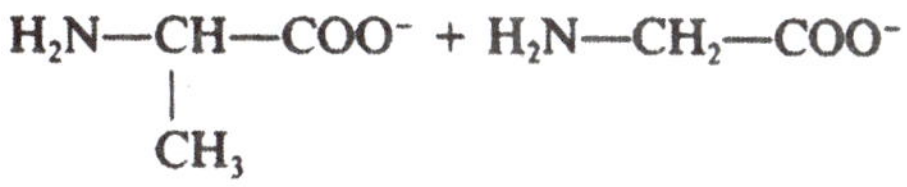

D. 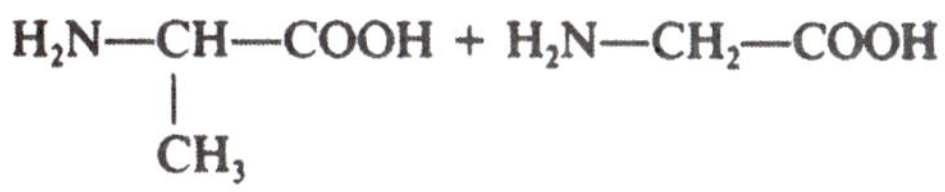 H$_2$N—CH(CH$_3$)—COOH + H$_2$N—CH$_2$—COOH

Passage II (Questions 9–14)

The functional derivatives of carboxylic acids are the acid chlorides, anhydrides, amides, and esters. These derivatives like carboxylic acids all contain the acyl group shown below and typically undergo nucleophilic substitution in which –OH, –Cl, –OOCR, –NH$_2$, or –OR' is replaced by some other basic group.

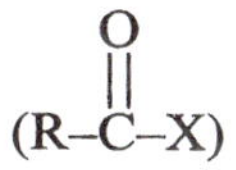

(R–C(=O)–X)

9. Aldehydes and ketones also contain the acyl group but they typically undergo addition reactions. The explanation for this difference is:

   A. the carbon of the carbonyl group in aldehydes and ketones is more electronegative than in the derivatives of carboxylic acids.
   B. the carbonyl carbon of carboxylic acid groups is sterically hindered.
   C. the X groups of the carboxylic acid derivatives are weaker bases than the hydride ion (: H$^-$) or alkide ion (: R$^-$) and are thus better leaving groups.
   D. the carbonyl carbon of aldehydes and ketones is sp^3 hybridized while the carbonyl of carboxylic acid derivatives has sp^2 hybridization.

10. The first step in nucleophilic acyl substitution reactions involves the formation of a tetrahedral intermediate:

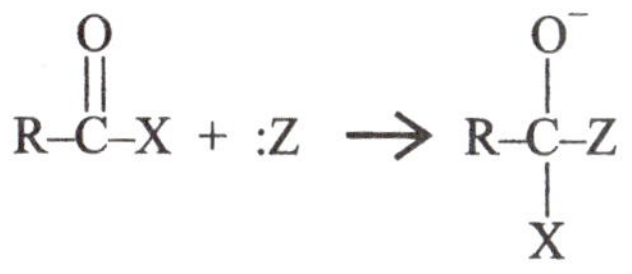

    R–C(=O)–X + :Z → R–C(O$^-$)(X)–Z

The rate of this reaction is increased by:

A. electron-withdrawing groups that stabilize the negative charge.
B. the presence of bulky groups.
C. Both A and B.
D. a neutral solution.

11. Nucleophilic substitution occurs much more readily at an acyl carbon than at a saturated carbon because:

    A. nucleophilic attack on a tetrahedral alkyl carbon gives rise to a crowded transition state.
    B. a pentavalent carbon is unstable.
    C. attachment of a nucleophile to a carbonyl carbon involves the breaking of a weak π bond.
    D. All of the above

12. The alcoholysis (cleavage by an alcohol) of an ester is called:

    A. acid hydrolysis.
    B. transesterification.
    C. hydrogeolysis.
    D. basic hydrolysis.

13. Mineral acids speed up esterification as well as the hydrolysis of esters. The best explanation is:

    A. there is a tetrahedral intermediate.
    B. the carbonyl carbon is protonated so that it is more susceptible to attack by a nucleophile.
    C. it gives rise to a better leaving group.
    D. the proton causes cleavage to occur between the oxygen and the acyl group.

14. Isotopic exchange experiments indicate that the greatest fraction of the tetrahedral intermediate that goes on to form product in the hydrolysis of carboxylic acid derivatives follows the sequences

    acid chloride > acid anhydride > ester > amide

    Which of the following is probably the most important factor?

    A. The nature of the solvent
    B. The acidity of the solution
    C. The basicity of the leaving group
    D. The basicity of the solution

CHAPTER 19
ENOLATE ANIONS AND ENAMINES (22)

Passage I (Questions 1–4)

Testosterone, a steroid androgen, or male sex hormone, is synthesized according to the following pathway.

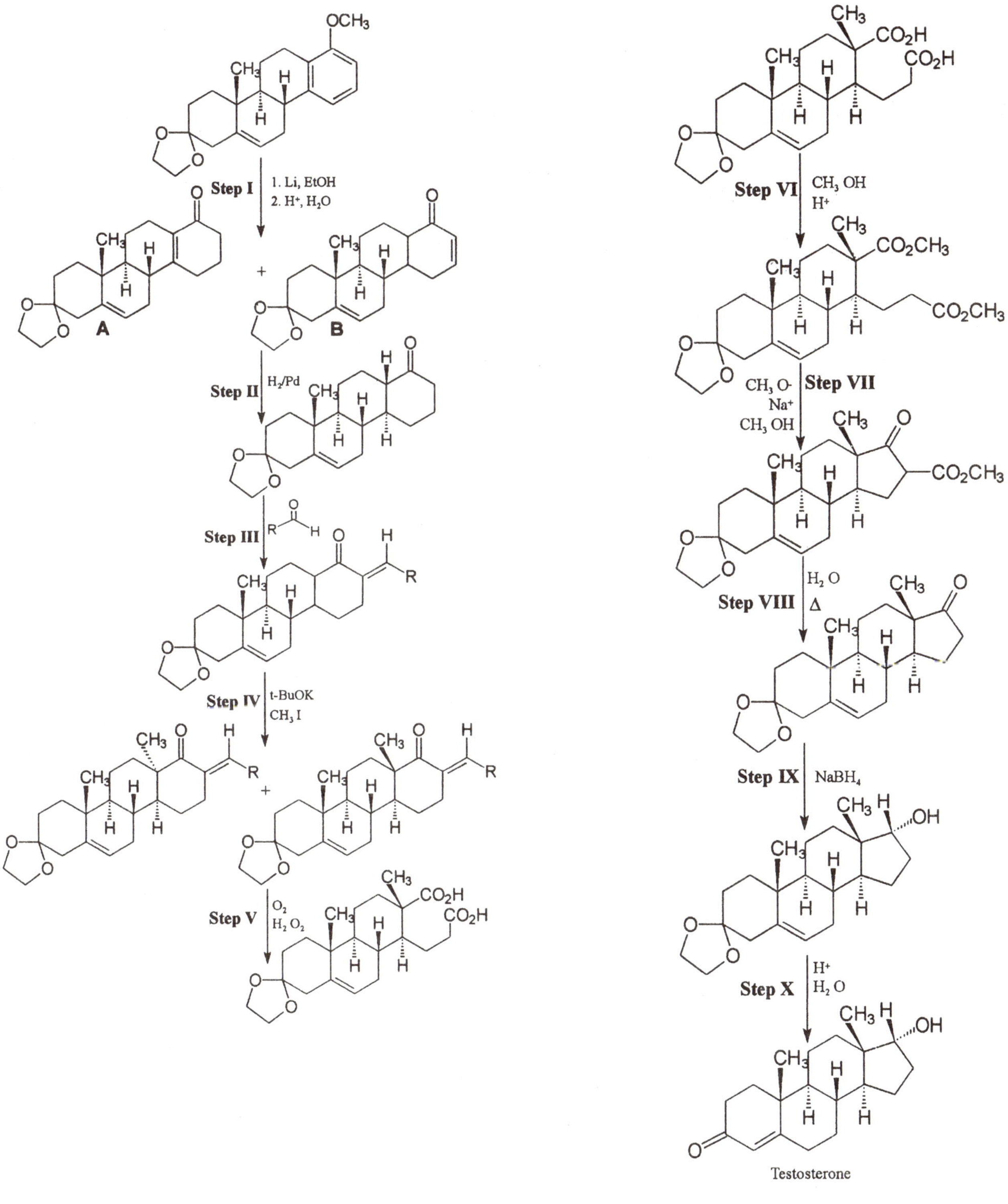

1. What is the product of the following reaction?

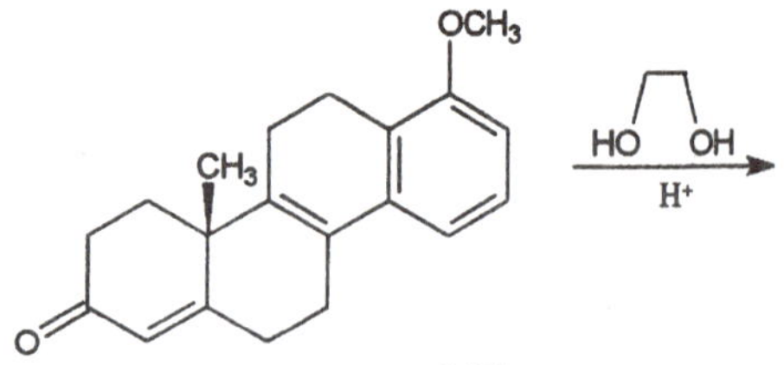

A.

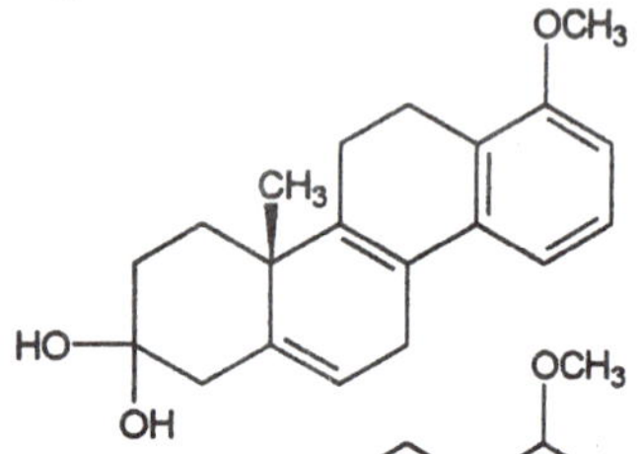

B.

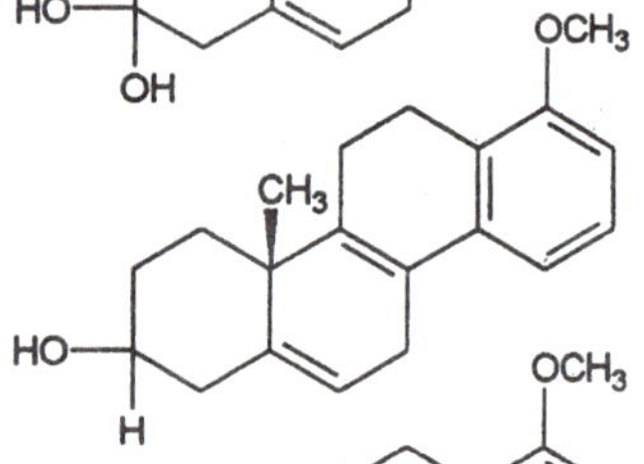

C.

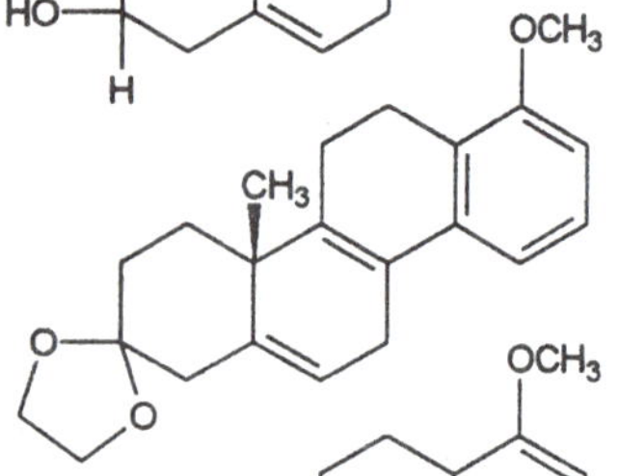

D.

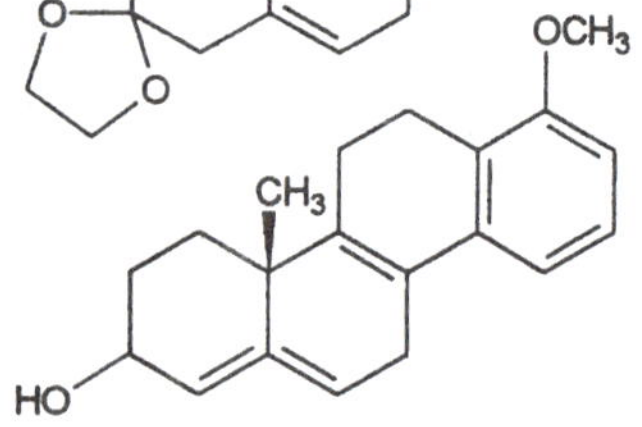

2. The synthesis of testosterone from the following precursor has been reported.

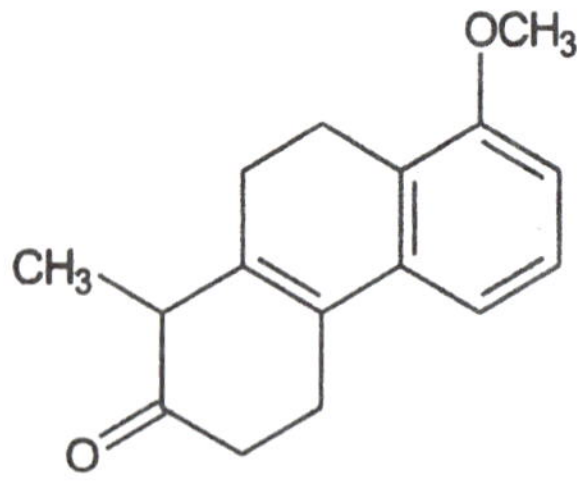

Which of the following is formed when a solution of the above compound is treated with lithium diisopropylamide (LDA); given that LDA is a hindered molecule whose conjugate acid has a pK_a of approximately 40?

A. A hemiacetal
B. An enolate ion
C. A radical anion
D. A ketal

3. Step 1 generated a mixture of cyclohexene derivatives A and B. Products A and B can most easily be distinguished from each other by ^{1}HNMR because:

   A. Product A has fewer vinylic protons than Product B.
   B. Product A has more vinylic protons than Product B.
   C. Product A has more protons than Product B.
   D. Product A has more sigma bonds than Product B.

4. In the synthesis of testosterone, Step IX is what type of reaction?

   A. A catalytic hydrogenation reaction
   B. A reduction reaction
   C. A saponification reaction
   D. A dehydration reaction

Passage II (Questions 5–10)

Tautomers are compounds that differ in their arrangement of atoms and undergo rapid equilibrium reactions. An example is enol-keto tautomerism:

$$-\overset{|}{C}=\overset{|}{C}-OH \rightleftharpoons -\overset{|}{\underset{H}{C}}-\overset{|}{C}=O$$

Because the equilibrium lies to the right, usually the reactions that produce the enol tautomer end up mostly with the keto form. The rearrangement takes place because of the polarity of the O-H bond, which enables H$^+$ to separate readily from oxygen:

$$-\overset{|}{C}=\overset{|}{C}-OH \rightleftharpoons -\overset{|}{C}=\overset{|}{C}-O^{\ominus}$$

$$\updownarrow$$

$$-\overset{|}{\underset{H}{C}}-\overset{|}{C}=O \rightleftharpoons -\overset{|}{C}-\overset{|}{C}=O + H^+$$

When the H$^+$ returns, it can join with the negatively charged carbon in the anion or go back to the oxygen. However, when H$^+$ attaches to carbon it tends to stay on much longer, favoring the keto tautomer.

5. Adding one mole of water to one mole of acetylene produces mostly:

   A. vinyl alcohol.
   B. a diol.
   C. acetaldehyde.
   D. acetone.

6. Which of the following statements best explains why the above equilibrium favors the keto form?

   A. Ketones are, in general, more stable than alcohols.
   B. Ketones are stronger acids.
   C. Equilibrium favors the weaker acid.
   D. Equilibrium favors the stronger acid.

7. Phenol is favored over its keto form.

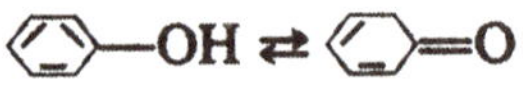

   ⟨⟩—OH ⇌ ⟨⟩=O

   Which of the following statements best explains this exception?

   A. Weaker acids tend to form from stronger acids.
   B. Extra stability associated with the aromatic phenyl ring makes phenol more stable.
   C. Oxygen does not have an octet in the keto form.
   D. Enols tend to be favored over their keto tautomers.

8. Another example of tautomerism is the enamine-imine equilibrium system:

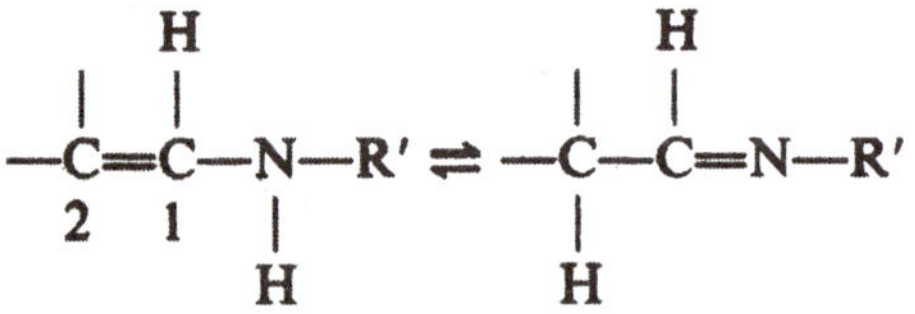

   $$-\underset{2}{C}=\underset{1}{C}-\underset{|}{N}-R' \rightleftharpoons -\underset{|}{C}-\underset{|}{C}=N-R'$$

   enamine   imine

   The acidic proton on the enamine form is attached to:

   A. R′
   B. C—2.
   C. N.
   D. C—1.

9. Carbonyl compounds react with 1° and 2° amines by nucleophilic addition. The addition product then undergoes dehydration. Which pair of reactions best describes these reactions?

   A. —C—C=O + RNH₂ → —C—C—N—R
              | | | |
              H   H OH H

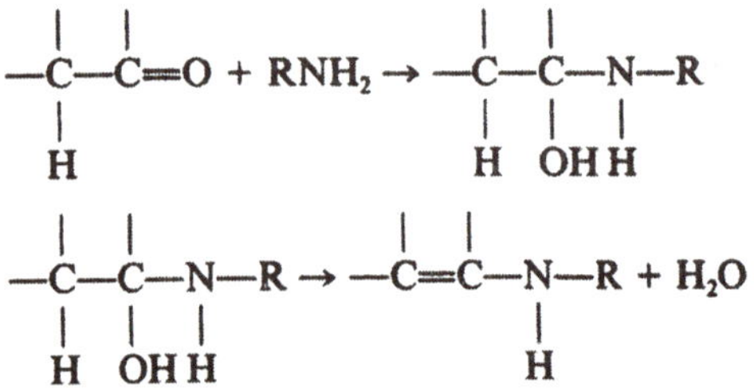

   —C—C—N—R → —C=C—N—R + H₂O
    | | | |
    H OH H   H

   B. —C=O + RNH₂ → —C—N—H
                       |
                     OR H

   —C—N—H → —C=NH + ROH
    |
    OR H

   C. —C—C=O + RNH₂ → —C—C—N—R
       | | | |
       H   H OH H

   —C—C—N—R → —C—C=NR + H₂O
    | | | |
    H OH H   H

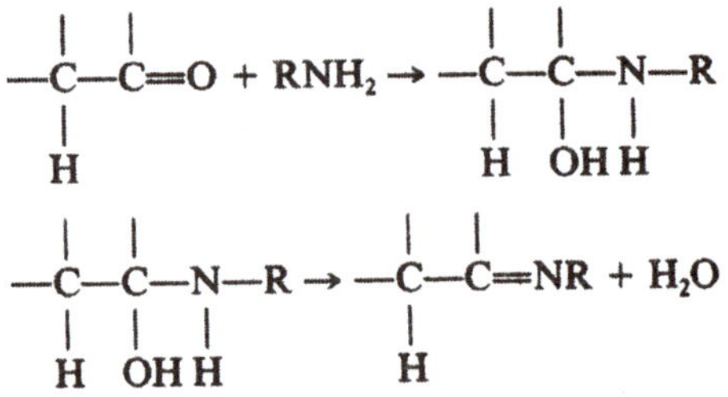

   D. —C—C=O + RNH₂ → —C—C—R
       | | |
       H   H OH

   —C—C—R → —C=C—R + H₂O
    | |
    H OH

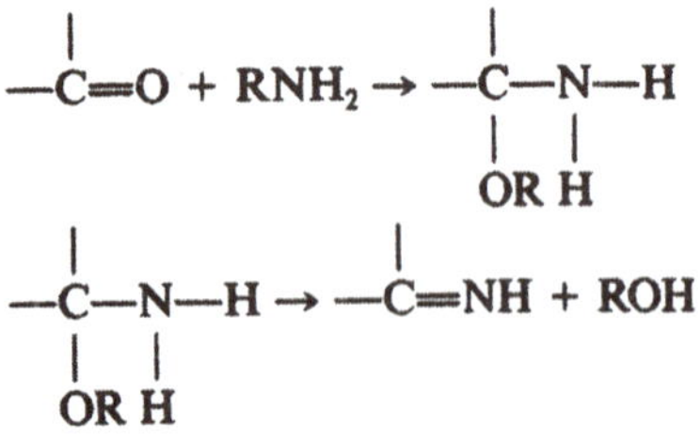

46

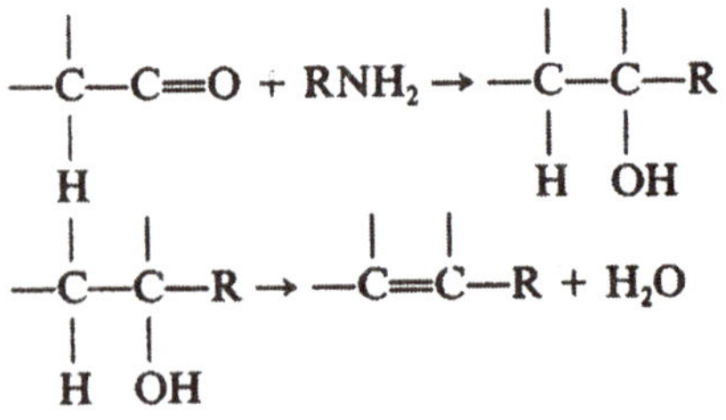

10. Which of the following enamines does not form an imine tautomer?

A. 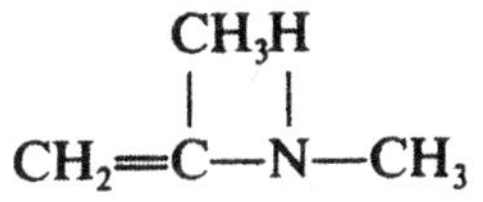 $CH_2{=}\underset{\underset{N(CH_3)(H)... }{}}{C}$ CH₂=C(CH₃)-N(H)(CH₃)

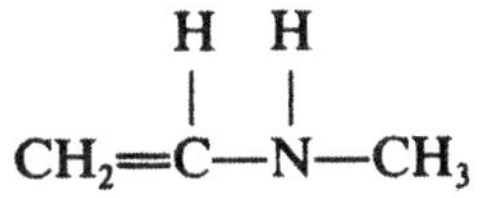

B. CH₂=C(H)-N(H)(CH₃)

C. H-C(CH₃)=C(CH₃)-N(H)(CH₃)

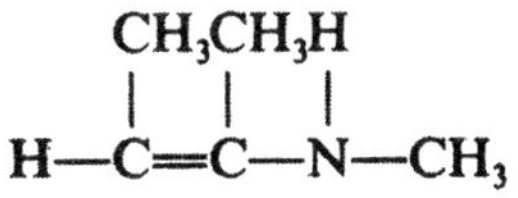

D. 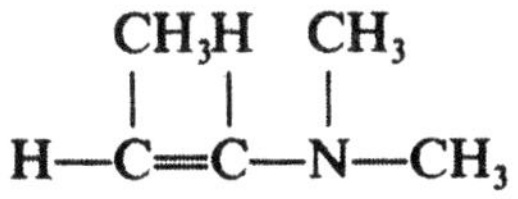 H-C(CH₃)=C(H)-N(CH₃)(CH₃)

11. In the presence of a base, two molecules of an aldehyde or a ketone can combine to form a β-hydroxyaldehyde or β-hydroxyketone in an aldol condensation. Which of the following compounds will not react in an aldol condensation?

A. 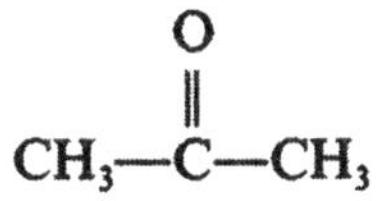 CH₃-C(=O)-CH₃

B. CH₃-C(=O)-CH₂-C(=O)-H

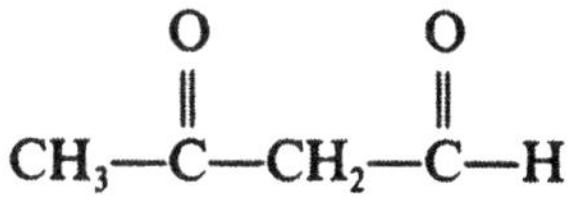

C. CH₃-C(CH₃)(CH₃)-C(=O)-H

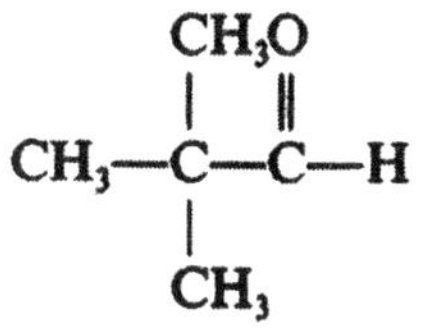

Passage III (Questions 11–15)

An organic chemistry student heats benzil with a strong base to form the benzilate anion:

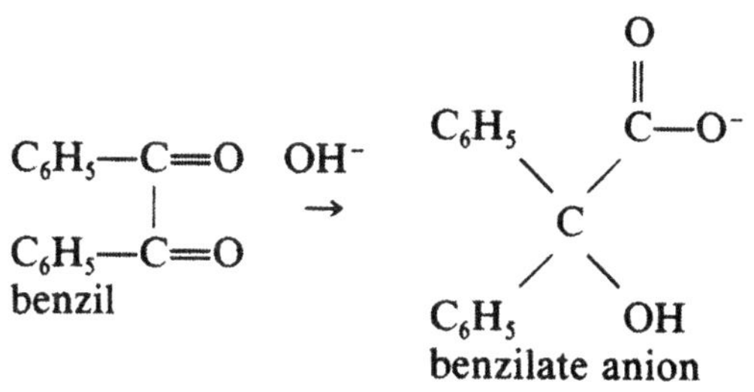

$C_6H_5-C(=O)-C(=O)-C_6H_5$ (benzil) + OH^- → $(C_6H_5)_2C(OH)-C(=O)-O^-$ (benzilate anion)

When methoxide ion in methanol is used instead of base, the methyl ester of benzilic acid is formed. Two mechanisms are considered:

47

Mechanism A.

$C_6H_5-C=O$ $\quad OH^-$ $\quad$ $C_6H_5-\overset{O^{\ominus}}{\underset{|}{C}}-OH$ $\quad$ $C_6H_5\diagdown \overset{\overset{O}{\|}}{\underset{\diagup \diagdown}{C}}-OH$ $\quad$ $C_6H_5\diagdown \overset{\overset{O}{\|}}{\underset{\diagup \diagdown}{C}}-O^-$
$C_6H_5-C=O$ $\quad \overset{\rightarrow}{\underset{\leftarrow}{}}$ $C_6H_5-C=O$ $\quad \overset{\rightarrow}{}$ $\quad C_6H_5 \quad O^-$ $\quad \overset{\rightarrow}{}$ $\quad C_6H_5 \quad OH$
$\quad \quad (1)$ fast $\quad (2)$ slow $\quad (3)$ fast

Mechanism B.

$C_6H_5-C=O$ $\quad OH^-$ $\quad C_6H_5-\overset{O^-}{\underset{|}{C}}-OH$ $\quad$ $C_6H_5\diagdown \overset{\overset{O}{\|}}{\underset{\diagup \diagdown}{C}}-O^-$
$C_6H_5-C=O$ $\quad \overset{\rightarrow}{(1)}$ $C_6H_5-\underset{|}{C}-OH$ $\quad \overset{\rightarrow}{(2)}$ $\quad C_6H_5 \quad O^-$ $\quad + H_2O$
$\quad \quad$ fast $\quad \quad O^-$ $\quad \quad$ slow

$C_6H_5\diagdown \overset{\overset{O}{\|}}{\underset{\diagup \diagdown}{C}}-O^-$
$\quad C_6H_5 \quad OH$ $\quad \overset{(3)}{fast}$

11. Both mechanisms involve a(n):

    A. reversible step.
    B. molecular rearrangement.
    C. nucleophilic substitution.
    D. electrophilic substitution.

12. If the methoxide ion in methanol is substituted for OH^-/H_2O, step (1) of Mechanism A will most likely yield which of the following?

    A. $C_6H_5-\overset{OCH_3}{\underset{|}{C}}=O$
       $C_6H_5-\underset{|}{C}=O$
       $\quad OCH_3$

    B. $C_6H_5-CH-OCH_3$
       $C_6H_5-CH-OCH_3$

    C. $C_6H_5-\overset{O^-}{\underset{|}{C}}-OCH_3$
       $C_6H_5-C=O$

    D. $C_6H_5-\overset{OCH_3}{\underset{|}{C}}=O$
       $C_6H_5-C=O$

48

13. Which of the following studies can be used to help determine the correct mechanism?

    A. Carry out the same reaction using O^{18}-labeled water.
    B. Measure the rate of formation of benzilate anion for different initial concentrations of benzil.
    C. Measure the rate of formation of benzilate anion for different pH values.
    D. Carry out the same reaction using a deuterated phenyl group.

14. Aliphatic diketones like the one illustrated below do not undergo similar reactions in alkaline solution because:

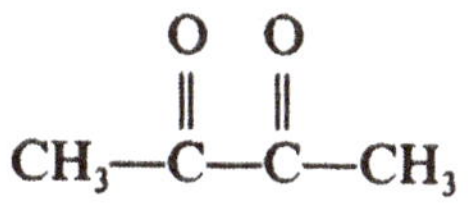

    $$CH_3-\overset{O}{\underset{\|}{C}}-\overset{O}{\underset{\|}{C}}-CH_3$$

    A. they never undergo nucleophilic addition reactions.
    B. they do not react with bases.
    C. they undergo condensation reactions due to the presence of α-hydrogen atoms.
    D. the central carbon-carbon bond undergoes cleavage in alkaline solution.

15. Which of the following reactions is unlikely?

    A. benzilic acid + CH_3NH_2 → amide formation
    B. benzilic acid + CH_3OH → esterification
    C. benzilic acid → CH_3COOH → acylation of alcohol group
    D. benzilic acid + HCN → substitution of –OH by –CN.

Passage III (Questions 16–20)

The reactivity of carbon-carbon double bonds toward nucleophilic addition is increased by the presence of an electron-withdrawing substituent. Thus α, β-unsaturated ketones, acids, esters, and nitriles undergo nucleophilic addition reactions that simple alkenes do not undergo. On the other hand, electron-withdrawing substituents deactivate the carbon-carbon double bond toward electrophilic addition as well as determine the orientation of the addition. (See Reaction 1 below.) The following reactions are known to occur:

Reaction 1.

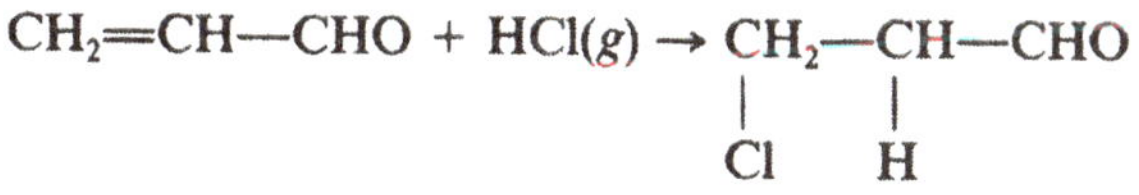

$$CH_2=CH-CHO + HCl(g) \rightarrow CH_2-CH-CHO$$
$$\hspace{4cm} |\hspace{0.5cm}|$$
$$\hspace{4cm} Cl\hspace{0.3cm} H$$

Reaction 2.

$$\text{Ph}-\underset{H}{\overset{H}{C}}=\underset{H}{\overset{H}{C}}-COOH + NH_2OH \rightarrow \text{Ph}-\underset{NHOH}{\overset{H}{C}}-\underset{H}{\overset{H}{C}}-COOH$$

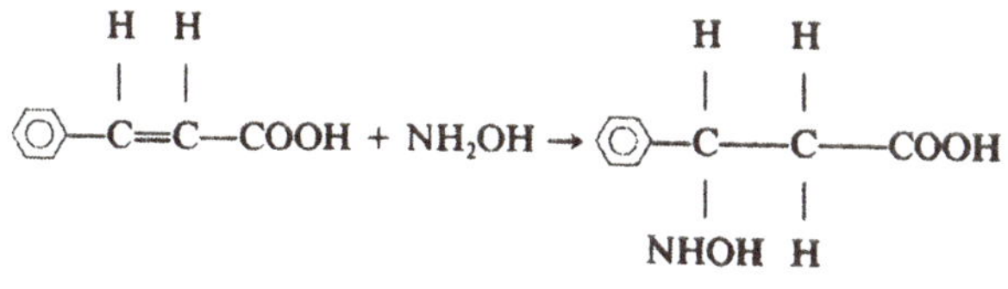

16. From the above reactions, it appears that a basic molecule or ion adds to α,β-unsaturated carbonyl compounds in the:

    A. α position.
    B. β-position.
    C. carbonyl oxygen.
    D. carbonyl carbon.

49

17. An H⁺ ion will most likely attack the α, β-unsaturated carbonyl compound at the:

    A. α carbon.
    B. β-carbon.
    C. carbonyl oxygen.
    D. carbonyl carbon.

18. What product(s) will be formed from the following reaction?

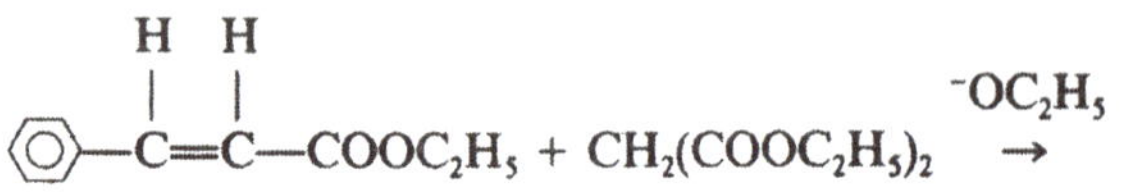

    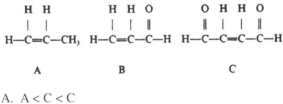

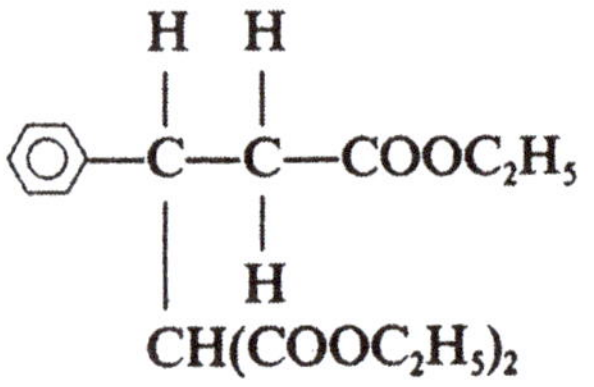

19. A plausible explanation for the fact that electron-withdrawing groups activate nucleophilic addition is that:

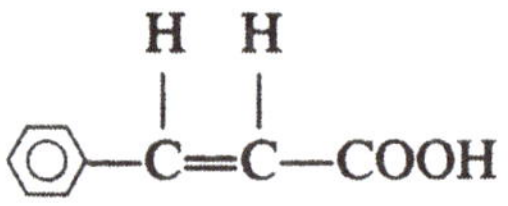

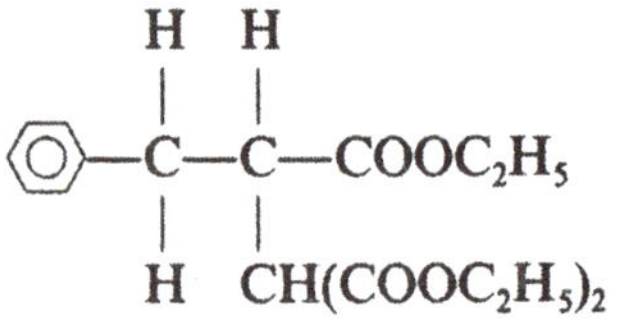

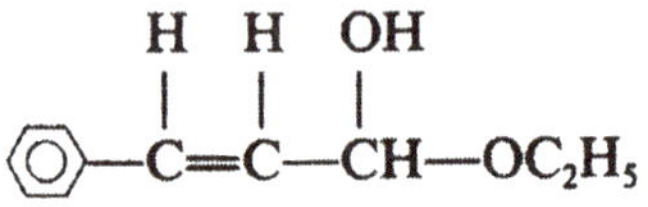

    A. they stabilize an anionic intermediate by dispersing the negative charge.
    B. they increase the negative charge on the carbon atoms of the double bond, increasing their susceptibility to attack by nucleophiles.
    C. they destabilize the transition state.
    D. they stabilize an anionic intermediate by dispersing the negative charge through inductive effects only.

20. Given compounds A, B, and C below, rank them in order of increasing reactivity toward :CN⁻

    H H   H H O   O H H O
    | | | | || || | | ||
    H—C=C—CH₃   H—C=C—C—H   H—C—C=C—C—H

         A   B   C

    A. A < C < C
    B. B < C < A
    C. A < B < C
    D. C < B < A

21. Acetone and acetaldehyde in the presence of a strong base will yield which of the following?

    A. Two different β-hydroxyaldehydes
    B. Four different β-hydroxycarbonyl compounds
    C. Two different β-hydroxyketones
    D. Three different β-hydroxycarbonyl compounds

CHAPTER 20
ATOMATICS I: BENZENE AND ITS DERIVATIVES (3)

1. Ortho-nitrophenol has a low boiling point and aqueous solubility due to:

    A. intermolecular H-bonding.
    B. intramolecular H-bonding.
    C. van der Waals forces.
    D. London dispersion forces.

2. Toluene will show peaks in its:

    A. IR spectrum only.
    B. UV and IR spectrum only.
    C. IR, UV, and NMR spectrum.
    D. UV spectrum only.

3. In keto-enol tautomerism:

    A. the enol form is usually more stable.
    B. phenol is an exception.
    C. the keto form is usually more stable.
    D. B and C are both true.

CHAPTER 21
ATOMATICS II: REACTIONS OF BENZENE AND ITS DERIVATIVES (1)

1. Benzene reacts with bromine to form bromobenzene, but the reaction requires the use of an appropriate Lewis acid catalyst such as ferric bromide.

    $$\text{C}_6\text{H}_6 + \text{Br}_2 \xrightarrow{\text{FeBr}_3} \text{C}_6\text{H}_5\text{Br} + \text{HBr}$$

    Reaction I

    It is reasonable to assume that Reaction I will proceed via which of the following reaction mechanisms?

    A. Nucleophilic substitution first order
    B. Nucleophilic substitution second order
    C. Electrophilic substitution first order
    D. Electrophilic substitution second order

CHAPTER 22
AMINES (17)

Passage I (Questions 1–5)

Figure 1 illustrates a solubility based characterization procedure, often used by organic chemists, for the qualitative analysis of monofunctional organic compounds.

Table 1 lists the organic compounds comprising the various solubility classes of Fig. 1.

Table 1

Group	Compounds
I	Salts of organic acids, amino acids, amine chlorides Sugars (carbohydrates) and other polyfunctional compounds with hydrophilic groups
II	Arenesulfonic acids Monofunctional carboxylic acids, alcohols, ketones, aldehydes, esters, amides, and nitriles with 5 or less carbon atoms Monofunctional amines with 6 or less carbon atoms
III	Phenols with ortho- and/or para-electron withdrawing groups, β-diketones Carboxylic acids with 6 or more carbon atoms
IV	Sulfonamides, nitro-compounds with α-hydrogens Phenols, oximes, enols, imides, and thiophenols with 6 or more carbon atoms
V	Some oxy-ethers, anilines, aliphatic amines with 8 or more carbon atoms
VI	Neutral compounds containing sulfur or nitrogen with 6 or more carbon atoms
VII	Ethers with 7 or less carbon atoms Monofunctional esters, aldehydes, ketones, cyclic ketones, methyl ketones with between 6 and 8 carbon atoms; epoxides
VIII	Ethers, most other ketones Unsaturated hydrocarbons, aromatic compounds, particularly those which possess activating groups

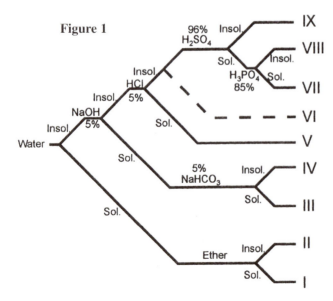

Figure 1

52

1. Phenols are soluble in a strongly basic sodium hydroxide solution, and insoluble in dilute sodium bicarbonate. Phenol has a pK_a = 10.0. The introduction of an ortho bromine atom into the phenol would have the effect of:

   A. lowering the pK_a and thus decreasing the acidity of the phenol.
   B. lowering the pK_a and thus increasing the acidity of the phenol.
   C. increasing the pK_a and thus decreasing the acidity of the phenol.
   D. increasing the pK_a and thus increasing the acidity of the phenol.

2. A certain water insoluble compound is insoluble in 5% sodium hydroxide, insoluble in 5% HCl, and insoluble in concentrated H_2SO_4 and insoluble in 5% sodium bicarbonate. In which class should this compound be classified?

   A. III
   B. IV
   C. VIII
   D. IX

3. Benzoic acid should be soluble in which of the following solvent pairs?

   A. Water and 5% HCl
   B. 5% NaOH and 5% $NaHCO_3$
   C. 5% HCl and 5% NaOH
   D. 85% H_3PO_4 and 5% NaOH

4. Low molecular weight amino acids fall into group I in Table 1. This is most likely due to the fact that low molecular weight amino acids are:

   A. hydrophilic and basic.
   B. hydrophobic and amphoteric.
   C. hydrophobic and lipophilic.
   D. hydrophilic and amphoteric.

5. A critical step in acid-catalyzed ester hydrolysis is the attack of water molecules on the protonated ester. If the water molecules are replaced with an alcohol, then the reaction will involve:

   A. transesterification of one ester to another.
   B. reduction to the corresponding aldehyde.
   C. base-protonated ester hydrolysis to the corresponding acid.
   D. decarboxylation to the corresponding C-H unit.

Passage II (Questions 6–11)

The ninhydrin reaction is a useful analytical detection method for α-amino acids. The reagent ninhydrin produces a characteristic blue color with primary α-amino acids via the following series of reactions (see Fig. 1):

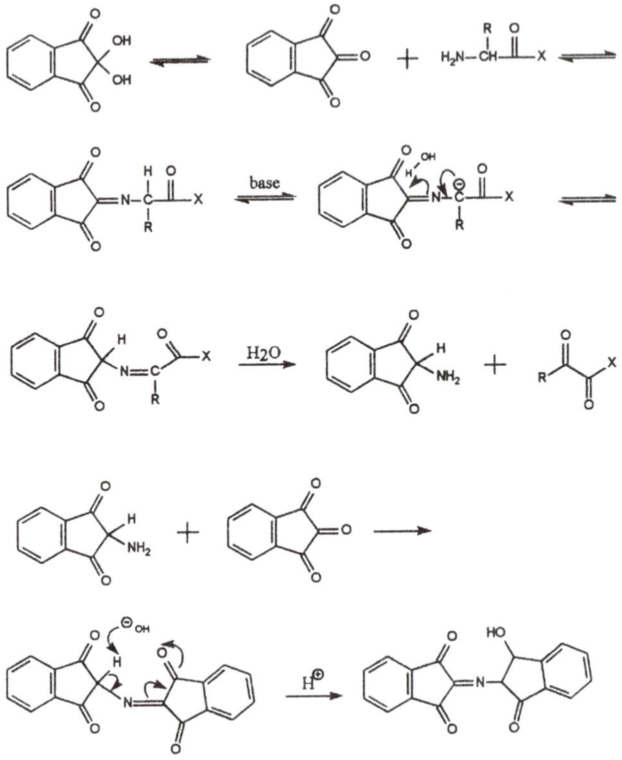

Figure 1

6. In the first reaction of Fig. 1, the inorganic product, which is not written, is:

   A. CO_2
   B. H_2O
   C. HCl
   D. H_2O_2

7. Which of the following compounds would be isotopically labeled if $H_2{}^{18}O$ were the only isotype source?

   A. 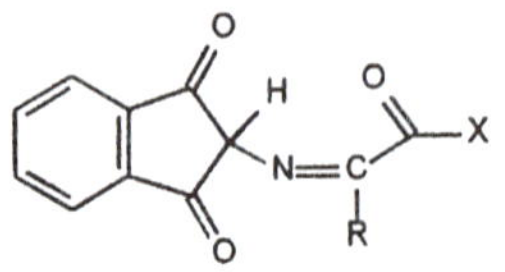
   B. 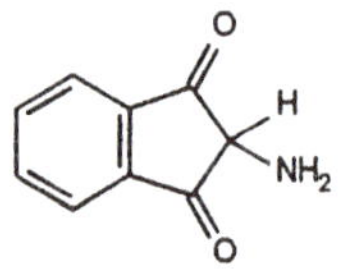
   C.
   D.

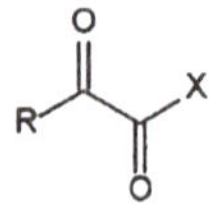

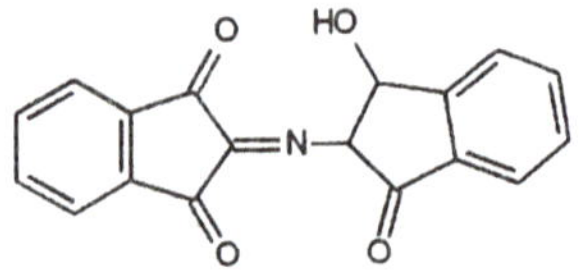

8. Base treatment of an amino acid usually results in the conversion of the acid to a derivative via the amino-carboxylate salt.

$$R-CH(NH_3^+)-CO_2^- \xrightarrow{B^-} R-CH(NH_2)-CO_2^- + BH$$

The above procedure:

A. decreases the rate of electrophilic reaction of the free amino group.
B. decreases the rate of nucleophilic reaction of the free amino group.
C. enhances the rate of nucleophilic reaction of the free amino group.
D. enhances the rate of electrophilic reaction of the free amino group.

9. A mixture of alanine and benzoyl chloride is treated with dilute aqueous sodium hydroxide to yield compound X. What functional group would be present in compound X?

A. Ester
B. Aldehyde
C. Amide
D. Ether

10. Amino acids can be divided into the following four general categories based on their acid-base charge properties at intracellular pH (~6-7):

  Positively charged
  Negatively charged
  Hydrophobic
  Hydrophilic

Consider the following amino acids.

I. 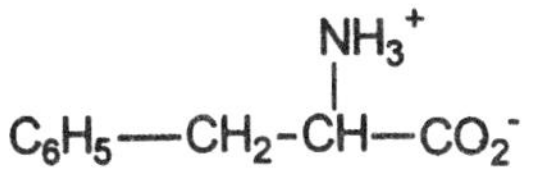 $C_6H_5-CH_2-CH(NH_3^+)-CO_2^-$

II. 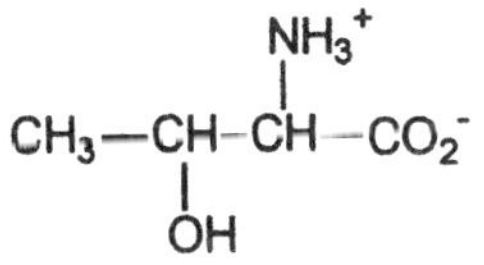 $CH_3-CH(OH)-CH(NH_3^+)-CO_2^-$

III. 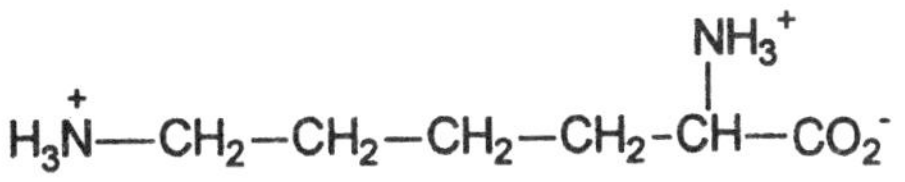 $H_3N^+-CH_2-CH_2-CH_2-CH_2-CH(NH_3^+)-CO_2^-$

IV. 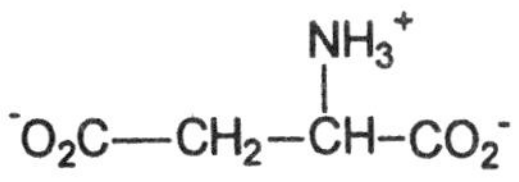 $^-O_2C-CH_2-CH(NH_3^+)-CO_2^-$

Which of the following classification series best represents amino acids I, II, III, and IV, respectively?

A. Hydrophobic, hydrophilic, positively charged, negatively charged
B. Hydrophobic, positively charged, hydrophilic, negatively charged
C. Hydrophilic, negatively charged, hydrophobic, positively charged
D. Positively charged, negatively charged, hydrophilic, hydrophobic

11. The structure of valine is shown below.

$$(CH_3)_2CH-\underset{\underset{NH_2}{|}}{CH}-COOH$$

In extremely basic solutions valine possesses 2 basic sites $-CO_2^-$ and $-NH_2$. Monoprotonation of such an alkaline solution would yield which of the following products?

A. $(CH_3)_2CH-\underset{\underset{NH_2}{|}}{CH}-CH_2OH$

B. $(CH_3)_2CH-\underset{\underset{NH_3^+}{|}}{CH}-COO^-$

C. $(CH_3)_2CH-\underset{\underset{NH_3^+}{|}}{CH}-COOH$

D. $(CH_3)_2CH-\underset{\underset{NH_2}{|}}{CH}-COOH$

Passage III (Questions 12–16)

Fig. 1 illustrates chemical methods often used by organic chemists for qualitative analysis of water insoluble unknowns. Table 1 lists characteristic chemical tests of organic compounds. For instance, sodium bicarbonate tests that are positive are indicative of a carboxylic acid.

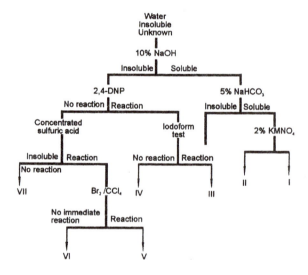

Figure 1

Table 1

Chemical Test	Compounds
Sodium hydroxide	Organic acids: carboxylic acids and phenols
Lucas	Alcohols with 5 or less carbon atoms
Sodium bicarbonate	Carboxylic acids
2,4-Dinitro-phenylhydrazine	Aldehydes and ketones
Iodine in sodium hydroxide (Iodoform)	Acetaldehydes and ketones with the CH_3-CO- group. Alcohols with the $CH_3CH(OH)$- as a structural feature
Sulfuric acid	Alcohols, ethers, alkenes. Soluble Lewis bases
Bromine	Alkenes

12. Heptane falls into group VII. This is primarily due to the fact that heptaine is:

    A. amphiphilic.
    B. hydrophobic.
    C. hydrophilic.
    D. amphoteric.

13. A certain water insoluble unknown is soluble in 10% NaOH, 5% $NaHCO_3$, and gives a negative test result for the Bayer test. Which of the following is the most probable structure of this unknown?

    A. 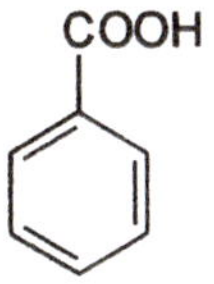 benzoic acid (COOH on benzene ring)

    B.  cyclohexene

    C. 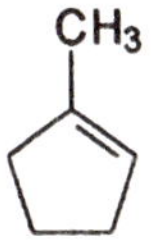 1-methylcyclopentene

    D. 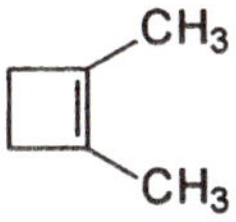 1,2-dimethylcyclobutene

14. The ^1HNMR of compound X was performed with $CDCl_3$ as a solvent and 8 signals were observed. A second ^1HNMR of compound X was performed after the addition of D_2O to the ^1HNMR solution, and seven signals were observed. The vanishing of the ^1HNMR signal is an indication that X contains which of the following functional groups?

    A. A carboxylic acid
    B. An acidic hydrogen
    C. An alkene
    D. An aldehyde

15. Consider the following aliphatic amines (Et = ethyl).

    I. Et$_3$N
    II. Et$_2$NCH$_2$CH$_2$F
    III. Et$_2$NCH$_2$CH$_2$Cl

    Which of the following series most accurately represents the basicity of these aliphatic amines in decreasing order?

    A. I, II, III
    B. II, III, I
    C. III, II, I
    D. I, III, II

16. A substance X is found to be soluble in sodium bicarbonate and has the molecular formula C$_3$H$_6$O$_2$. The IR spectrum has demonstrated a band from 3000 cm^{-1} to 2500 cm^{-1} and a sharp absorption peak at 1715 cm^{-1}. Based on the preceding data, compound X is most likely which of the following?

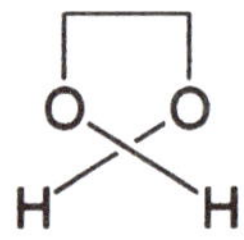

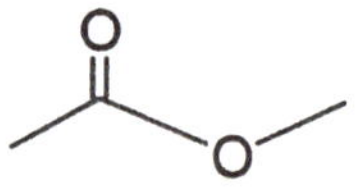

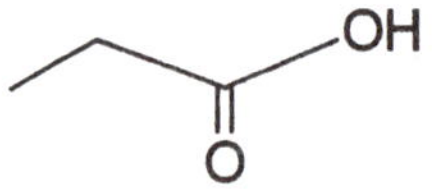

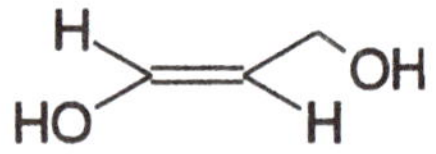

17. The dipolar ion of alanine is

    $$\overset{\oplus}{H_3N}-CH(CH_3)-COO^{\ominus}$$

    What is the predominant form in basic solution?

    A. H$_2$N—CH(CH$_3$)—COOH

    B. $\overset{\oplus}{H_3N}$—CH(CH$_3$)—COOH

    C. $\overset{\oplus}{H_3N}$—CH(CH$_3$)—COO$^{\ominus}$

    D. H$_2$N—CH(CH$_3$)—COO$^{\ominus}$

CHAPTER 23
CONJUGATED SYSTEMS (10)

Passage I (Questions 1–5)

Conjugated dienes undergo many reactions typical of alkenes such as electrophilic addition. Electrophilic additions invariably produce a mixture of products which result from competing pathways.

For example, 1,3-butadiene reacts with 1 equivalent of Br_2 at 40 °C to give dibromide products A and B, in 20% and 80% yield, respectively. The addition reaction occurs via a resonance stabilized allylic intermediate.

$$CH_2 = CH - CH = CH_2 + Br_2 \rightarrow [\text{intermediate}] \rightarrow A + B$$
Reaction I

When the reaction is carried out at –80 °C, the product composition changes to 80% A and 20% B. Product A of the mixture is easily isolated to its pure state by recrystallization at 54 °C.

Both products A and B are stable at –80 °C.

$$A \text{ or } B \xrightarrow{-80°C} \text{no change}$$

In contrast, upon studying in an ionizing solvent at 40 °C, either product is converted to a mixture identical to that produced by the addition of 1 equivalent of Br_2 to 1,3-butadiene at 40 °C.

$$A \text{ or } B \xrightarrow{40°C} \underset{(20\%)}{A} + \underset{(80\%)}{B}$$

1. Which of the following pairs most accurately depicts the resonance stabilized intermediate that led to products A and B?

   I. $CH_2=CH-\overset{-}{C}HCH_2Br \leftrightarrow \overset{+}{C}H_2-CH=CHCH_2Br$

   II. $CH_2=CH-\overset{+}{C}HCH_2Br \leftrightarrow \overset{+}{C}H_2-CH=CHCH_2Br$

   III. $CH_2=CH-\overset{+}{C}HCH_2Br \leftrightarrow \overset{-}{C}H- CH=CHCH_2Br$

   A. I only
   B. II only
   C. I and II only
   D. II and III only

2. If the bromination of 1,3 butadiene is carried out at –80 °C the major product is Compound A. However, if the reaction is carried out at 40 °C Compound B becomes the major product. Which of the following statements most accurately explains this transformation?

   A. A is the thermodynamic product and B is the kinetic product.
   B. A is the kinetic product and B is the thermodynamic product.
   C. A has a disubstituted double bond and is somewhat less stable than B which has a monosubstituted double bond.
   D. A has a monosubstituted double bond and is more stable than B which has a disubstituted double bond.

3. What is the IUPAC name of product A?

   A. 3,4-dibromo-1-butene
   B. 1,4-dibromo-1-butene
   C. 1,4-dibromo-2-butene
   D. 3,4-dibromo-2-butene

4. Since products A and B have the same molecular formula (C$_4$H$_6$Br$_2$), it is reasonable to assume that products A and B:

   A. are enantiomers.
   B. form a racemic mixture.
   C. are structural isomers.
   D. are conjugated dienes.

5. The addition of bromine to cyclohexene will result in which of the following products?

   A.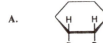

   B.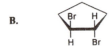

   C.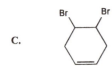

   D.

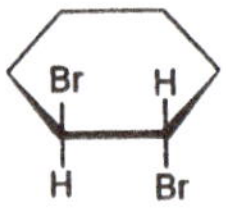

Passage II (Questions 6–10)

An organic chemist is interested in studying free radical addition to conjugated dienes. The following reactions are performed:

Experiment 1.

  Step (1) Peroxide decomposes to form a free radical peroxide → R•.

  Step (2) The free radical abstracts bromine from BrCCl$_3$

  R• + BrCCl$_3$ → R — Br + •CCl$_3$

  Step (3) The •CCl$_3$ adds to the conjugated system

  •CCl$_3$ + CH$_2$=CH—CH=CH$_2$ → allylic free radical

  Step (4) allylic free radical

  allylic free radical $\xrightarrow{\text{BrCCl}_3}$ Cl$_3$C—CH$_2$—CH—CH=CH$_2$ +
  $\quad\quad\quad\quad\quad\quad\quad\quad\quad$ |
  $\quad\quad\quad\quad\quad\quad\quad\quad\quad$ Br

  Cl$_3$C—CH$_2$—CH=CH—CH$_2$—Br

Experiment 2.

  BrCCl$_3$ is also reacted with a 50:50 mixture of 1,3-butadiene and 1-octene. BrCCl$_3$ reacts mostly with 1,3-butadiene.

6. Based on the above reactions, which of the following statements is false?

   A. Conjugated dienes undergo addition by free radicals.
   B. Alkenes undergo addition by free radicals.
   C. Simple alkenes undergo free radical addition reactions faster than conjugated dienes.
   D. Conjugated dienes undergo free radical addition reactions faster than simple alkenes.

7. Which radicals are formed by the addition of •CCl$_2$ to 1,3-butadiene?

   A. Cl$_3$C—CH$_2$—ĊH—CH=CH$_2$

       Cl$_3$C—CH$_2$—CH=CH—ĊH$_2$

   B. Cl$_3$C—ĊH—CH$_2$—CH=CH$_2$

       Cl$_3$C—CH$_2$—ĊH—CH=CH$_2$

   C. Cl$_3$C—CH$_2$—CH=ĊH—CH$_2$

       Cl$_3$C—ĊH—CH$_2$—CH=CH$_2$

   D. Cl$_3$C—CH$_2$—ĊH—CH=CH$_2$

       Cl$_3$C—ĊH—CH=CH—CH$_3$

8. Which statement is supported by Experiment 2?

   A. The transition state of the diene is more stable than the transition state of the simple alkene.
   B. The diene is more stable than the simple alkene.
   C. The activation energy for the diene is lower than the activation energy for the simple alkene.
   D. The activation energy for the diene is higher than the activation energy for the simple alkene.

9. Both alkenes and dienes undergo free radical polymerization. Polymers formed from dienes, however, differ from polymers of alkenes in that polymers formed from dienes:

   A. contain double bonds.
   B. are saturated.
   C. require an initiator to begin the polymerization.
   D. are all substituted.

10. Dienes like simple alkenes undergo electrophilic addition. Butadiene when treated with bromine forms which of the following?

    A. 3,4-dibromo-1-butene only
    B. 1,4-dibromo-2-butene only
    C. 3,4-dibromo-1-butene and 1,4-dibromo-2-butene
    D. 1-bromo-2-butene only

CHAPTER 25
CARBOHYDRATES (9)

1. The structure of D-glucose is shown below:

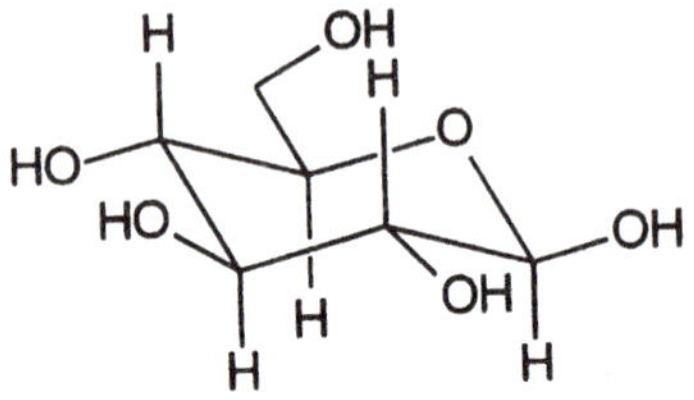

   An alternative projection system, often used in carbohydrate chemistry is the Fischer system. Fischer projections are particularly useful devices for depicting the absolute configuration at the various stereocenters of a compound. Which of the following most accurately represents the modified Fischer projection of D-glucose?

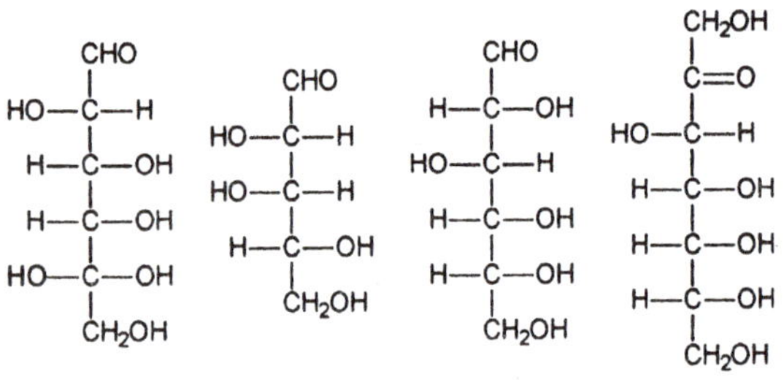

   A. B. C. D.

2. The structure of β-D-glucose is shown below in two different projection sytems. The circled hydroxyl group in Fig. 1 would be located at which position in the modified Fischer projection depicted in Fig. 2?

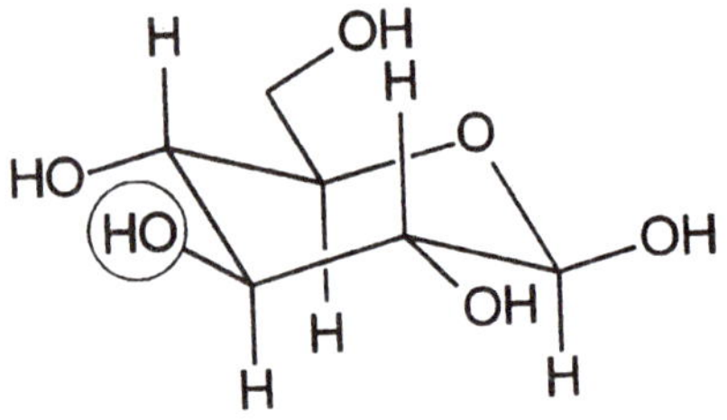

   Figure 1

   A. I
   B. II
   C. III
   D. IV

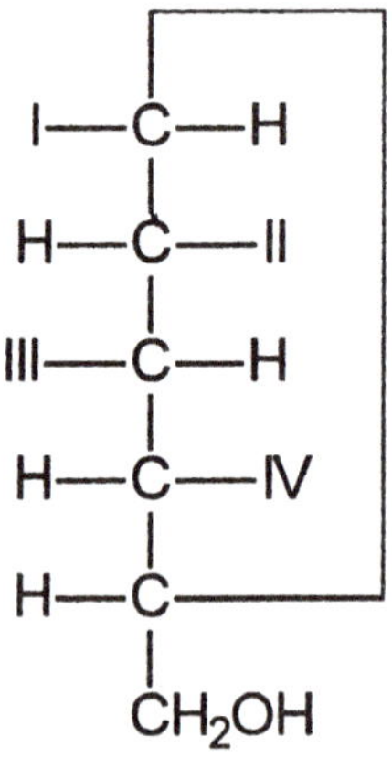

   Figure 2

3. The α and β forms of D-glucose are:

   A. enantiomers.
   B. epimers.
   C. meso structures.
   D. anomers.

Passage I (Questions 4–9)

Starch, cellulose, and glycogen are all made of (+)-glucose, making it the most abundant monosaccharide in nature. The first structure proposed for D-(+)-glucose, a pentahydroxy aldehyde was

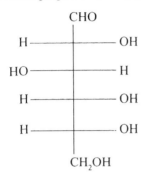

4. Experimental data which were inconsistent with this structure include:

   A. D-(+)-Glucose does not form a bisulfite addition product.
   B. D-(+)-Glucose undergoes osazone formation.
   C. D-(+)-Glucose exists in two isomeric forms which undergo mutarotation.
   D. Both A and C.

5. Aldehydes react with alcohols to form acetals according to

   $$\underset{H}{-C=O} \xrightarrow{CH_3OH/H+} \underset{\underset{OH}{|}}{\overset{\overset{H}{|}}{-C-OCH_3}} \xrightarrow{CH_3OH/H+} \underset{\underset{OCH_3}{|}}{\overset{\overset{H}{|}}{-C-OCH_3}}$$

   When D-(+)-glucose is treated with methanol and HCl, only one –CH$_3$ group is added although the product has the properties of a full acetal. The best explanation is that:

   A. glucose is not a pentahydroxy aldehyde.
   B. glucose is not optically active.
   C. glucose has a cyclic structure.
   D. glucose exists in two isomeric forms.

6. Which of the following most closely describes glucose in aqueous solution?

   A. A straight chain structure that is in equilibrium with a cyclic structure
   B. Two planar cyclic diastereomers differing in configuration about C-1
   C. A six-membered, puckered, pyranose ring
   D. A six-membered pyranose ring in a chair conformation

7. β-D-(+)-glucose is the most abundant isomer of naturally occurring glucose because:

   A. it is not stable.
   B. every bulky group occupies an equatorial position in the chair conformation.
   C. most of the bulky groups occupy equatorial positions in the chair conformation.
   D. Both A and B.

8. Both cellulose and starch are made up of chains of D-glucose units joined by:

   A. double bonds.
   B. an ether linkage.
   C. C-1.
   D. glycosidic linkage.

9. D-glucose and D-mannose are examples of:

   A. stereoisomers.
   B. epimers.
   C. ketohexoses.
   D. fats.

CHAPTER 26
LIPIDS

1. The alcohol produced upon basic or acidic hydrolysis of a fat is which of the following?

   A. CH_2-CH_2 with OH, OH

   B. $CH_3-CH_2CH_2OH$

   C. $CH_3-CH-CH_2$ with OH, OH

   D. $CH_2-CH-CH_2$ with OH, OH, OH

2. Phosphoglycerides differ from fats because:

   A. phosphoglycerides have only two acyl groups while fats have three.
   B. fats are esters while phosphoglycerides are acid anhydrides.
   C. fats can be hydrolyzed while phospholipids cannot.
   D. phosphoglycerides have R groups while fats do not.

CHAPTER 27
AMINO ACIDS AND PROTEINS

1. Which of the following represents the amino acid methionine at its isoelectric point?

    A. 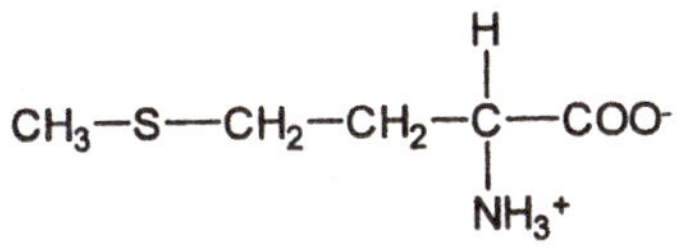 CH₃—S—CH₂—CH₂—C(H)(NH₃⁺)—COO⁻

    B. CH₃—S—CH₂—CH₂—C(H)(NH₃⁺)—COOH

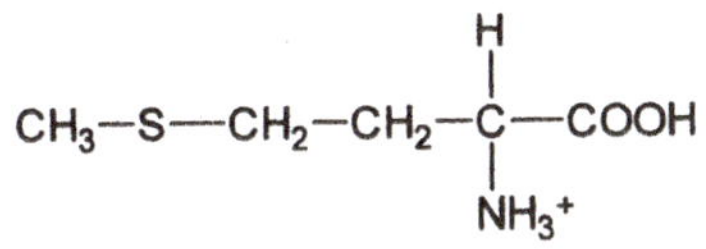

    C. CH₄⁺—S—CH₂—CH₂—C(NH₂)—COOH

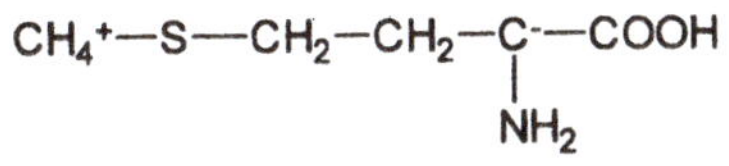

    D. 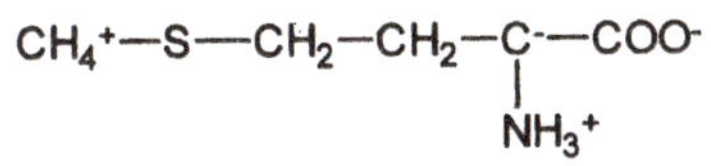 CH₄⁺—S—CH₂—CH₂—C(NH₃⁺)—COO⁻

Passage I (Questions 2–7)

A student titrates one liter of a 1.0 M amino acid solution, which is acidified with HCl with a NaOH solution. He plots pH versus moles of NaOH added to give the following graph:

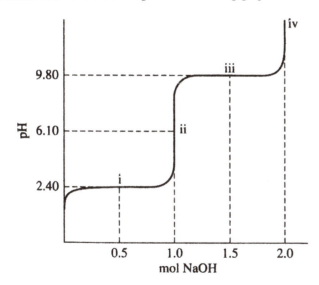

2. The titration curve indicates that the amino acid is:

    A. triprotic.
    B. diprotic.
    C. a base.
    D. monoprotic.

65

3. The amino acid can be identified by comparing:

   A. the first full neutralization point with pK_{a1}.
   B. the second full neutralization point with pK_{a2}.
   C. the isoelectric point (p*I*) with pK_{a1}.
   D. the pK_{a1} and pK_{a2} of the amino acid with the pH after addition of 0.5 mol and 1.5 mol of NaOH.

4. At which isoelectric point (pH) does the amino acid have a dipolar or zwitterioric form?

   A. 6.10
   B. 2.40
   C. 7.00
   D. 9.80

5. Which of the following is the predominant form of the amino acid after 0.60 mol of NaOH is added?

   A. 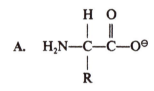   C.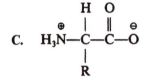

   B. 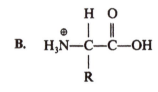   D.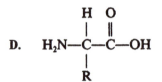

6. After 2.0 mol of NaOH is added, the pH of the solution is:

   A. higher than p*I*.
   B. lower than p*I* but lower than pK_{a2}.
   C. higher than pK_{a2}.
   D. lower than pK_{a1}.

7. A buffer solution is produced after the addition of how much NaOH?

   A. 1.0 mol
   B. 1.5 mol
   C. 2.0 mol
   D. 2.5 mol

8. How many amide linkages are in Ala-Gly-Phe?

   A. 1
   B. 3
   C. 0
   D. 2

9. Which of the following is correct?

   A. All amines react with acid chlorides to form amides.
   B. All peptides have amide linkages.
   C. Alkylation of a tertiary amine produces an amide.
   D. The alkylation of amines by alkyl halides is an electrophilic substitution reaction.

MCAT Answers And Explanations

Chapter 3

Answers

Q. 1 The correct answer is (A).

Q. 2 The correct answer is (B).

Q. 3 The correct answer is (A).

Q. 4 The correct answer is (C).

Answer Explanations

Q 1 This question shows the structure of 1,1-dichloroethane. The molecule's central carbon has two identical substituents (C1). Thus is it *not* chiral. Only chiral substances can rotate plane polarized light.

Q 2 This question tests your ability to identify a molecule which has no enantiomer. In other words, a molecule whose mirror image can be superimposed on itself. Examine molecule I. Note that its mirror image is molecule III. If you just picked up molecule III and tried to put it on top of molecule I, it doesn't work (*the hydroxyl groups never cover each other*). Even if you rotate molecule III around 180°, you still cannot superimpose it on molecule I. Molecules I and II are enantiomers of each other!

Examine molecule II. Draw its mirror image. Note that when you turn the mirror image 180°, you end up with the same molecule. Thus it has no enantiomer and may be referred to as a *meso* compound.

Q 3 tests your memory of the definition of enantiomers: they come in pairs and are two *nonsuperimposable* mirror images of each other. Answer choice A. is the only compound which you can draw the mirror image then twist the image to superimpose onto the original molecule.

Chapter 6

Answers

Q. 1 The correct answer is (C).

Q. 2 The correct answer is (B).

Q. 3 The correct answer is (C).

Q. 4 The correct answer is (A).

Q. 5 The correct answer is (D).

Answer Explanations

1. According to Markovnikov's Rule, when an acid, HX, adds to a double bond, the hydrogen atom bonds to the carbon atom that already is attached to the most hydrogen atoms. Thus the H in HBr bonds to C-1 and the Br to C-2 of 1-butene.

2. In this case, the H adds to C-2 to produce the more stable cation $H_2\overset{+}{C}–CH_2–CF_3$. Applying Markovnikov's Rule would produce the less stable

    $$H_3C-\underset{H}{\overset{+}{C}}-CF_3$$

    in which an electron-withdrawing group –CF_3 is bonded to an electron deficient carbon atom. Such a species is of high energy and therefore does not form to any extent.

3. If one applies Markovnikov's Rule then the usual carbocation would form. However carbocations sometimes rearrange to give a more stable species. In this case a methide ion ($:CH_3^-$) would shift to yield the more stable carbocation.

4. The addition of HX to alkenes proceeds through an open carbocation. With a p orbital at the electron-deficient carbon, the electrophile can attack equally as well from the top or bottom. Thus, the enantiomers are produced in equal amounts creating an optically inactive, racemic solution.

5. When the alkene is protonated the activated complex will resemble the carbocation intermediate. Because the primary carbocation is less stable, its associated activation energy will be greater and its rate constant smaller according to the Arrhenius equation: $k = Ae^{-E_a/RT}$. The tertiary carbocation will thus form more readily and lead to more product.

6. The correct answer is (C).

    $$CH_3-CH_2-\underset{\underset{CH_3}{|}}{CH}-CH_2^{\oplus} \xrightarrow{-H^+} CH_3-CH_2-\underset{\underset{CH_3}{|}}{C}=CH_2$$

    $$CH_3-CH_2-\underset{\underset{\oplus}{}}{\overset{\overset{CH_3}{|}}{C}}-CH_3 \xrightarrow{-H^+} CH_3-CH=\underset{\underset{CH_3}{|}}{C}-CH_3$$

7. The correct answer is (C). Loss of water or dehydration to form an alkane indicates an elimination reaction. The first order kinetics indicates an E1 elimination with the rate of the reaction dependent on the rate of formation of the carbocation intermediate.

8. The correct answer is (A).

    $$CH_3-CH=CH-CH_3 \xrightarrow{Cl_2} CH_3-\underset{\underset{Cl}{|}}{CH}-\underset{\underset{Cl}{|}}{\overset{\overset{CH_3}{|}}{C}}-CH_3$$

9. The correct answer is (B).

    The $CH_3-CH_2-\underset{\underset{CH_3}{|}}{CH}-CH_2^{\oplus}$ carbocation rearranges by a 1,2 hydride shift to form the more stable

CH₃—CH₂—C⁺(CH₃)—CH₃ with CH₃ branch, a tertiary carbocation. This rearrangement accounts for the chief product,

CH₃—CH=C(CH₃)—CH₃ which gives mostly 2,3-dichloro-2-methylbutane upon chlorination (Experiment 1). Experiment 4 shows a decrease in formation of the carbocation due to deuterium being substituted for hydrogen. The C—D bond will break more slowly, hindering the 1,2 hydride shift.

10. The correct answer is (B).

CH₃—CH₂—CH₂—CH(CH₃)—CH₂Br → CH₃—CH₂—CH₂—C⁺(CH₃)—CH₃ → CH₃—CH₂—CH=C(CH₃)—CH₃
                                         chief carbocation

11. The correct answer is (B).

12. The correct answer is (B). (S,S)-2,3-dibromobutane

13. The correct answer is (A). meso-2,3-dibromobutane

14. The correct answer is (D). Mechanism A as shown produces (S,S)-2,3-dibromobutane. Rotation about the C—C bond produces the meso compound. Bromination of cis-2-butene, however, yields a racemic mixture, so Mechanism A is incorrect. Mechanism B showing a bromonium intermediate produces equal concentrations of the enantiomers.

15. The correct answer is (B). The carbon carrying the positive charge of the carbocation has only six electrons.

16. The correct answer is (C). Mechanism A produces a very stable carbocation in which the positive charge is delocalized into the ring.

17. The correct answer is (B). As already indicated, only Mechanism B accounts for the stereochemistry of the bromination of the cis isomer. This is an anti addition since bromine attacks from opposite sides of the molecule. Anti addition must also occur in the bromination of the trans isomer in order to produce the meso product.

18. The correct answer is (A). Using Mechanism B

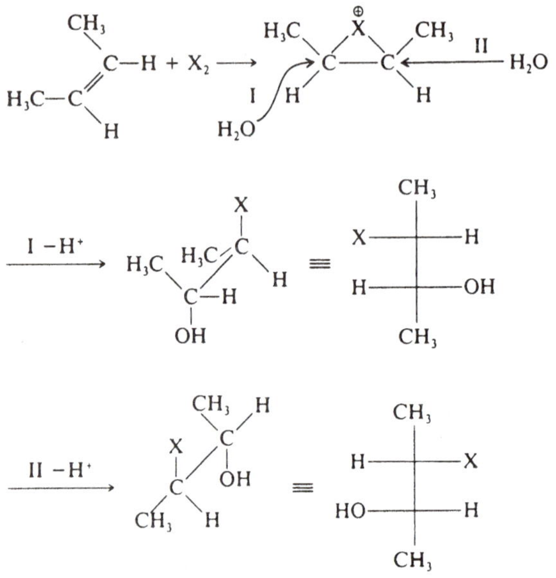

19. The correct answer is (B).

20. The correct answer is (D). It is nonsuperimposible on its mirror image.

Chapter 8

Answers

Q. 1. The correct answer is (D).

Q. 2. The correct answer is (B).

Q. 3. The correct answer is (C).

Q. 4. The correct answer is (D).

Q. 5. The correct answer is (D).

Q. 6. The correct answer is (C).

Q. 7. The correct answer is (B).

Q. 8. The correct answer is (C).

Answer Explanation

Since this is our first organic chemistry passage together we'll discuss it in detail. Afterwards, you'll begin to notice the trends then we'll speed up a bit more.

Principle #1: Organic Chemistry is simple.

Principle #2: It's OK to learn new words (i.e. *solvolysis*) during the exam!

Let's solve the problem. Read the first paragraph. Now you should want to know what tert-butyl bromide looks like. Draw it below.

Principle #3: Think about the preceding molecule. Begin by assessing electronics: bromine is a halogen to the right of carbon on the periodic table, thus it is more electronegative (inscribe δ next to Br). Since Br is drawing electrons away from the central carbon which it is attached to, that carbon becomes partially positive, δ^+.

Principle #4: Recognize. Notice how bulky the molecule is. Compare the molecule with less hindered ones with the same number of carbons, i.e., sec-butyl which has a secondary carbon or n-butyl (*normal* butyl) which has a primary carbon. Steric hindrance means that the partially positive carbon has so many attached groups around it that if you were a nucleophile, with your negative or partially negative charge, you would have a difficult time gaining access to that carbon which attracts you. If you can't reach the carbon *nucleus* then you can't engage in *nucleo*philic substitution.

Remember, you're still a nucleophile. You want something positive but you can't get to the carbocation. However, you have your pick of hydrogens! Removing or *eliminating* H from t-butyl is the start of an *elimination* reaction. Note, for this reaction to occur, we would need a hydrogen-hungry strong nucleophile like OH$^-$ or $C_2H_5O^-$ (*ethoxide*, which was added to Reaction I in the passage).

KEY CONCEPT: If we started with a primary compound (i.e. n-butyl bromide) and added a nucleophile (you!), the nucleophile has easy access, and is attracted, to the partially positive carbon. Thus the nucleophile adds to the central carbon and bumps off the leaving group (Br) = *nucleophilic substitution, second order* = S_N2. "Second order" emphasizes that the rate-determining step depends on the concentration of two compounds — the nucleophile and n-butyl bromide.

However, this problem begins with a tertiary hindered compound which, as we describe, may undergo *elimination*, and for the same reason as previous, it is also *second order* = E2. Once a hydrogen is eliminated by the nucleophile, we are left with the very unstable but intermediate primary carbanion whose electrons are quickly put to use by forming a double bond and bumping off Br$^-$. Thus the product is $(CH_3)_2C=CH_2$. Its name is 2-methyl-1-propene. It contains hydrogens that are arranged similar to those of the vinyl group. Draw the E2 mechanism and the product below.

Is there a second possible mechanism? That question is both excellent and rhetorical! Here is an expression to memorize: *tertiary carbocations are relatively stable*. Why? Here is another expression to memorize: *alkyl groups are somewhat electron donating*. In other words, if you were a primary carbocation, you would be pretty unstable! You would try to hold on to a formal positive charge without any help—too difficult. However, if you were a tertiary carbocation surrounded by three alkyl groups—each of which will tend to reduce the burning positive charge on the central carbon with cool electrons!

Because of the stability of tertiary compounds, the secondary mechanism has one*** compound in its rate-determining step: t-butyl bromide in solution can simply dissociate into Br$^-$ and the stable t-butyl$^+$. Draw this reaction.

Now a nucleophile would be happy to *quickly* mate with the positive carbocation (*nucleophilic substitution,* first *** order = S_N1). If the nucleophile is hydroxide, or water, then the product would be the tertiary alcohol tert-butanol $(CH_3)_3C$-OH *note* —OH *substituted* —Br. If the nucleophile is ethoxide, or ethanol, then the product would be the ether $(CH_3)_3C$-OCH_2CH_3).

Now let's read the second paragraph in the problem! By analyzing the possible mechanisms we now can say with confidence that Compound A must be the S_N1 product ethoxy t-butane [$C_6H_{14}O = (CH_3)_3C-OCH_2CH_3$]. Compound B, with its vinylic protons, must be the E2 product 2-methyl-1-propene [$C_4H_8 = (CH_3)_2C=CH_2$]. The latter is the answer to **Q 2**.

The third paragraph tells us what we already know! This includes: (i) there are two possible mechanisms each of which are identified with different rate constants k_1 and k_2. (ii) since the rate is given by:

$$\text{rate} = k_1 [\text{t-BuBr}] = k_2 [\text{t-BuBr}][C_2H_5O^-]$$

Thus the first reaction mechanism (k_1) is first order (S_N1) while the second (k_2) must be second order (E2). {*the order of the reaction = the sum of exponents*}

Back to the passage! After some irrelevant info, we come to the final two paragraphs concerning Reaction II. Once again, we already KNOW ALL THIS STUFF!!! To begin with, the ethanol in the solvent system created the compounds A and B according to the S_N1 and E2 mechanisms we've discussed. Then Compound C is produced by the reaction of t-BuBr and the water in the solvent system producing tert-butanol ($CH_3)_3C-OH$ (S_N1 which we looked at a few paragraphs ago. . .). Thus the passage gives us the correct molecular formula, $C_4H_{10}O$, and the correct IR absorbance for an alcohol (= hydroxyl group, **Q 3**).

Q 5 can be answered by looking at the rate expression or by working out the S_N1 mechanism as we did. Note that E2 is not an answer choice. But if it were, it still would not be the best answer! "Pure ethanol" = no ethoxide = no hydrogen-hungry strong nucleophile = E2 is about as probable as adoring your in-laws. It's not impossible, but it's not likely either.

Q 6 reminds us that the S_N1 mechanism occurred in *both* reactions I and II. In other words, t-Bu$^+$ is the stable tertiary carbocation intermediate prior to the nucleophilic reaction with ethanol/ethoxide (Reaction I) or water (Reaction II).

Q 7 has been answered many different ways already: A= S_N1, B=E2, and C- S_N1.

Q 8 points to the second paragraph of the passage which suggests that the reaction occurred spontaneously, which means $\Delta G < 0$, which also means that the great likelihood is that $\Delta H < 0$. The latter is called an *exothermic* reaction. Since energy is released, the reactants must have a higher energy and the products must have a lower energy. The only possible answers are A. and C. However, the intermediate in A. has a low energy which indicates stability implying that any further reaction is not likely to be spontaneous (from the mechanisms we did we know this to be false, untrue and unpleasant to hear). Answer choice C. suggests a higher energy intermediate which would be happy to engage in a further reaction to create a low energy, very stable final product.

Answers

Q. 9. The correct answer is (A).

Q. 10. The correct answer is (C).

Q. 11. The correct answer is (C).

Q. 12. The correct answer is (C).

Answer Explanations

This is a fun passage {☺!} Begin by drawing the molecular structures of isopropyl bromide, Na$^+$ ethoxide:

The ethoxide ion, being relatively small and negatively charged, will (*but* not exclusively) be attracted to the δ^+ central carbon in isopropyl bromide. Thus ethoxide is the nucleophile and Br is the leaving group = S_N2. The product would be ethyl isopropyl ether. Draw this S_N2 mechanism:

However, with the big bulky base, Na⁺ tert-butoxide, it would have difficulty accessing the central carbon in isopropyl bromide. Instead, it takes the easy way out: it plucks protons off a molecule like oranges off a tree! {I think I've said that before!} This leaves the unstable primary carbanion intermediate whose electrons are quickly attracted to the neighboring δ⁺ carbon thus bumping out Br = E2 mechanism. Thus the product is propene; *yes, Na⁺ ethoxide often engages the E2 reactions!! However, the difference between the two oxides in this problem is* size *which affects the* probability *of the occurrence of a given reaction*). Draw the E2 mechanism for the production of propene:

In the second paragraph in the passage, we learn what we have already worked out! Compound A is propene, C_3H_6, the reaction occurs by *second* order kinetics (=E2), and it has vinylic protons. In the third paragraph we also learn that its molecular mass is 42 grams.

Paragraph 3 introduces Compound B which we already worked out as $C_5H_{12}O$, ethyl isopropyl ether.

Paragraphs 4 and 5 are irrelevant!

Q 9 has been answered in the preceding paragraphs (Compound A = E2, B = S_N2).

Q 10 has been worked out to reveal Compound A as prop*ene*, an alk*ene*.

Q 11 relies on your understanding of the key events in the E2 mechanism as illustrated in ORG 6.2.4. The dotted lines are meant to represent *partial* bonds. Thus we're looking for an option that shows the attraction of the electrons from a negatively charged group (i.e. ethoxide) to a proton; as the *positive* proton is being *eliminated* (*E2*) the *negative* electrons on the primary carbon are being attracted to the neighboring δ⁺ carbon with the good leaving group which is electronegative. Answer choice C. fills our requirements! Answer choice D. represents the S_N2 mechanism.

Q 12 was determined to be $C_5H_{12}O$, ethyl isopropyl ether.

Answers

Q. 13. The correct answer is (B).

Q. 14. The correct answer is (B).

Q. 15. The correct answer is (B)

Q. 16. The correct answer is (A).

Q. 17. The correct answer is (B).

Q. 18. The correct answer is (D).

Answer Explanations

Yeah! More mechanisms! Let's make some alkenes . . .

Experiment 1 uses isopropyl iodide (= an alk*yl* hal*ide* - 2-iodopropane) and $K^+\,^-OCH_2CH_3$ (*recall: alkoxides are strong bases*, in a classic E2 reaction. By telling you that the reaction is first order with respect to each of the *two* reagents simply emphasizes the fact that the reaction is second order *overall*, E2.

Q 13 is answered by the above. Draw the mechanism and E2 product below. The product is propene (= *propylene*).

The above represents the '94% yield'. For fun, draw below how the *minor* product could be formed through either of the two nucleophilic substitution mechanisms. Either way, this minor product is called ethyl isopropyl ether.

Q 14 is so simple it makes me sad (O). Draw propylene which you created in the preceding mechanism and add the two hydrogens from H_2.

Q 15 tests your respect for $KMnO_4$ as a powerful agent. It can add oxygen(s) to reactants. Under abrasive conditions it can cleave the double bond; however, 'cold and dilute' is not abrasive. Thus the product is ethylene glycol.

Q 16 turns back to the passage. In Experiment II, 1-bromopentane is treated with the strong base KOH. The possible mechanisms to explain the products which they tell us are produced are either: (i) OH^- removes a proton from C-2 (= *the 2nd carbon in 1-bromopentane*) creating a secondary carbanion intermediate which produces 1-pentene via E2 (12% yield); or (ii) ethanol or ethoxide (*which could be produced in low yield from* ethanol + OH^-) simply attacks the δ^+ C-1 kicking out or *substituting* -Br with $-OCH_2CH_3$ creating ethyl pentyl ether via S_N2 (88% yield).

Q 17 asks us to remember 'Mark's rule': alkene + acid $\Rightarrow$ under ionic conditions (i.e. no: uv, ht, ↑ energy/heat), hydrogen adds preferentially to where its buddies are (= *the greatest number of other H's at the double bond* = a simplification of Markovnikoff's rule).

Q 18 is an extension of the previous answer. Reaction I abides by Mark's rule producing the Markovnikoff product 2-bromobutane, preferentially. Reaction II uses the same reactants but via a free radical mechanism. By definition, the anti-Markovnikoff product is produced preferentially (1-bromobutane).

Chapter 9

Answer

1. The correct answer is (C).

Answer Explanation

1. This question tests your understanding of a catalyst: they speed up the rate of reaction, they decrease the activation energy, they do not affect K_{eq}, they are not used up in a reaction, and finally, to make a reaction more inclined to occur, ΔG, which is a measure of spontaneity, *decreases*.

Answer

2. The correct answer is (D).

Answer Explanation

2. The correct answer is (D).

Chapter 13

Answers

Q. 1. The correct answer is (D).

Q. 2. The correct answer is (B).

Answer Explanations

Q 1 tests your memory of a couple of the IR absorption peaks. The absolute minimum to memorize are the bands for an alcohol (OH, 3200 - 3650) and that for the carbonyl group (C=O, 1630 - 1780) because these are the two most encountered functional groups in MCAT organic chemistry.

Bottle II has a peak at 1710 (carbonyl) + 3333 - 3500 (hydroxyl) = carboxylic acid (i.e. benzoic acid). Bottle IV has a peak at 3333 so it must be the alcohol. Without doing anything else, there is only one possible answer, D.

Q 2 will see if you have the most basic understanding of a proton NMR. Not including the reference peak at zero, there is only one peak! This means two things: (i) each H must be living in an environment which is identical to any other H in the entire molecule (only answer choice B. is possible); and (ii) since there is only one peak, to calculate the number of H's *n* on an adjacent carbon: $n + 1 = 1$. Thus *n* is zero indicating that each H is attached to a carbon whose neighboring carbon has no attached H's (again, only option B. is possible).

Answers

Q. 3. The correct answer is (C).

Q. 4. The correct answer is (A).

Q. 5. The correct answer is (B).

Q. 6. The correct answer is (A).

Q. 7. The correct answer is (D).

Answer Explanations

3. $CH_3-\underset{\underset{Br}{|}}{\overset{\overset{CH_3}{|}}{C}}-CH_2Br$

There are two kinds of protons in this molecule: the six methyl protons and the two —CH_2— protons. Splitting is not observed because these non-equivalent protons are not on adjacent carbons. Hence, two NMR signals are observed in a 3:1 ratio.

4. There are four non-equivalent protons: methyl protons (c), —CH₂— protons (d), and the two vinyl protons (a) and (b). The vinyl protons are non-equivalent because of their different chemical environments.

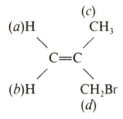

Consequently, 4 NMR signals will be observed.

5. A deuteron gives no signal in a proton NMR spectrum because it absorbs at a much higher field. Consequently, by replacing H with D, signals are removed from the spectrum so that only two NMR signals will appear in a 3:2 peak ratio. No splitting will occur because the non-equivalent hydrogens are not attached to adjacent carbon atoms.

```
   D         CH₃
    \       /
     C = C
    /       \
   D         CH₂Br
```

6. The Br atom withdraws negative charge via an inductive effect deshielding the proton and causing a downfield shift.

7. Nuclei of some elements generate a magnetic moment. When such nuclei are exposed to a magnetic field, their magnetic moment can be aligned with or against the field. The magnetic moment of the nucleus is aligned with the field in the more stable state. Energy can be absorbed to excite the nucleus into a higher energy state in which the magnetic moment of the nucleus is aligned against the field. An NMR spectrum records these absorptions. Some examples of nuclei which generate a magnetic field include protons, ^{13}C, and ^{19}F.

Answers

Q. 8. The correct answer is (D).

Q. 9. The correct answer is (A).

Q. 10. The correct answer is (C).

Q. 11. The correct answer is (B).

Q. 12. The correct answer is (C).

Answer Explanations

8. 3350 cm^{-1} —OH stretch (H—bonded) 1000 cm^{-1} C—O stretch
According to the table provided, the above two bonds are associated with alcohols.

9. The ultraviolet spectrum of an unknown compound is used chiefly to identify conjugation or the presence of an aromatic group.

10. Proton exchange occurs very rapidly in the presence of an acid or base. Consequently, the hydroxyl proton sees only an averaged chemical environment and thus gives rise to a singlet. If proton exchange is slowed, the hydroxyl proton signal will be split by nearby protons. The hydroxyl proton will also cause splitting in the signals of these nearby protons.

12. IR is the best method to detect a $-\overset{|}{C}=O$ group. The strong band due to C = O stretching is generally seen at about 1700 cm^{-1} where it seldom is hidden by other absorptions.

Chapter 16

Answers

Q. 1. The correct answer is (C).

Q. 2. The correct answer is (A).

Answer Explanations

Q 1 begins with your ability to write the molecular structure for formaldehyde (*a carbonyl group with two hydrogens attached to carbon*). Next, let's treat formaldehyde with water. Draw the mechanism below.

The δ^- oxygen in water is attracted to the δ^+ carbonyl carbon. The pi bond breaks and the free electrons go to oxygen. The attached water molecule now has three bonds (two H's and one carbon) and thus is positively charged. Electronegative oxygen is not happy! Oxygen pulls electrons to itself from its bond to H, thereby kicking off the proton H$^+$. The proton is attracted to the oxygen with the free electron pair. Our product is a *diol* (= 2 alcoh*ol* groups). Note that the mechanism is the same as for hemiacetal formation (ORG 7.2.2), where R=H, R'=H, and R'''=H.

Q 2

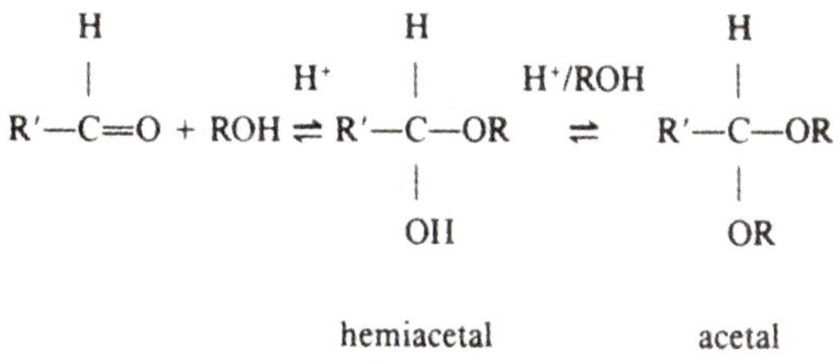

$$R'-C(H)=O + ROH \underset{H^+}{\rightleftharpoons} R'-\underset{OH}{\overset{H}{\underset{|}{C}}}-OR \underset{H^+/ROH}{\rightleftharpoons} R'-\underset{OR}{\overset{H}{\underset{|}{C}}}-OR$$

hemiacetal   acetal

Chapter 18

Answers

Q. 1. The correct answer is (A).

Q. 2. The correct answer is (B).

Q. 3. The correct answer is (C).

Q. 4. The correct answer is (A).

Q. 5. The correct answer is (D).

Q. 6. The correct answer is (A).

Answer Explanations

1. This bond must be broken in order to account for the labeled alcohol oxygen appearing in the ester.

2. The mechanism of acidic hydrolysis must be the exact reverse of the mechanism of esterification catalyzed by acid. If a labeled oxygen of an alcohol appears in the ester, hydrolysis of the ester will form the labeled alcohol back again.

3. Protonation of an ester yields:

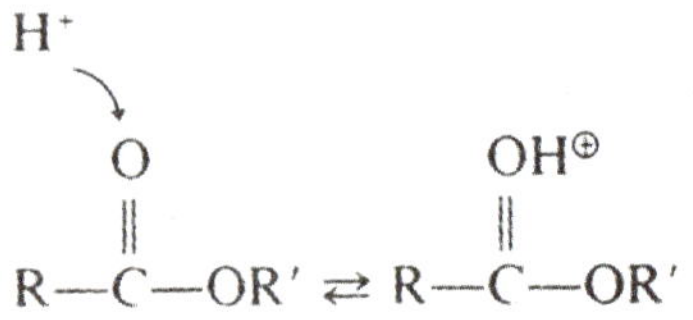

$$\underset{\text{R—C—OR'}}{\overset{\overset{H^+}{\searrow}\;\;\;\overset{O}{\|}}{}} \rightleftarrows \underset{\text{R—C—OR'}}{\overset{\overset{OH^{\oplus}}{\|}}{}}$$

The carbon is then rendered more susceptible to attack by a nucleophile because the oxygen will not need to carry a negative charge. The nucleophile in acidic hydrolysis is H_2O while in esterification the alcohol is the nucleophile.

4. Special methods are often needed to prepare esters of tertiary alcohols or esters of acids containing bulky groups.

5. $\underset{\text{acyl group}}{R-\overset{\overset{O}{\|}}{C}\!\!\not{\,\,}^{18}OC_2H_5}$

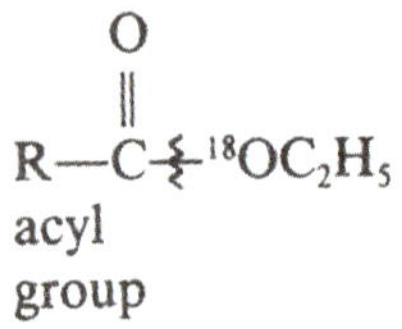

Answers

Q. 7. The correct answer is (D).

Q. 8. The correct answer is (C).

Answer Explanations

7. Quaternary nitrogen atoms cannot react to form an amide because it already has four bonds.

8. Hydrolysis breaks the amide linkage to form the amino acid residues alanine and glycine. Because the acids are in basic solution, their dibasic form predominates.

Answers

Q. 9. The correct answer is (C).

Q. 10. The correct answer is (A).

Q. 11. The correct answer is (D).

Q. 12. The correct answer is (B).

Q. 13. The correct answer is (B).

Q. 14. The correct answer is (C).

Answer Explanations

9. For both classes of acyl compounds, the first step involves the addition of a nucleophile (:Nu) that forms a tetrahedral intermediate:

$$R-\overset{O}{\underset{\|}{C}}-X + :Nu \rightarrow R-\overset{O^-}{\underset{X}{\overset{|}{C}}}-Nu$$

$$R-\overset{O}{\underset{\|}{C}}-R + :Nu \rightarrow R-\overset{O^-}{\underset{R}{\overset{|}{C}}}-Nu$$

The ability of X to leave depends on its basicity: the weaker the base the better the leaving group. In the case of an aldehyde or ketone, the very strong bases, the hydride ion (:H⁻) or alkide ion (:R⁻) would have to leave in order for substitution to occur. Thus, the carboxylic acids tend to undergo substitution reactions while aldehydes and ketones tend to undergo addition reactions.

10. The tetrahedral intermediate is favored by electron withdrawing groups that delocalize the negative charge, and hindered by bulky groups that cause crowding.

11. Nucleophillic attack on a tetrahedral alkyl carbon generates a crowded transition state and partially broken sigma bond.

12. In the hydrolysis of an ester, an alcohol is displaced by a nucleophile. In the same way, one alcohol can displace another alcolhol from an ester. This cleavage by an alcohol (alcoholysis) of an ester is called transesterification. Transesterification is catalyzed by acid or base.

13. Both electronic factors (the tendency of oxygen to acquire electrons) and steric factors (relatively unhindered transition state associated with the trigonal acyl reactant becoming the tetrahedral intermediate) make the carbonyl group susceptible to nucleophilic attack at the carbonyl carbon. In the presence of acid, H⁺ attaches to the carbonyl oxygen, making the carbonyl carbon even more susceptible to attack by a nucleophile: Oxygen can acquire the π electrons of the carbonyl group without developing a negative charge.

14. The basicity of the leaving group: $Cl^- < R'COO^- < OR'^- < NH_2^-$

Chapter 19

Answers

Q. 1. The correct answer is (C).

Q. 2. The correct answer is (B).

Q. 3. The correct answer is (A).

Q. 4. The correct answer is (B).

Answer Explanations

No, you are not expected to know the mechanisms of synthesis of the millions of biochemicals in nature! Yes, you are supposed to remain cool under pressure. No, you are not expected to be intimidated by the mere SIZE of a molecule! And yes, you should know your mechanisms well enough such that if different molecules are presented to you, you can key in on the *functional groups* you know to be reactive.

Q 1 provides us with only one answer which could possibly have the correct geometry! Nonetheless, let's work through the mechanism.

The story goes something like this: The catalyst (H^+), being the most charged substance, is implicated first. Thus the electrons from the electronegative oxygen (O in the carbonyl, C=O) are attracted to the proton (H^+) and bonds. To remain neutral oxygen loses its pi bond with carbon, leaving only a single bond and secondary carbocation. The δ^- charge on the oxygen from the *diol* (= a compound with 2 alcohol — OH — groups) attacks the positively charged carbocation. The extra hydrogen on the oxygen which now attaches to carbon is kicked out as a proton (regenerating our catalyst). Now we have our "*hemi-ketal*": the ketone has been converted into a hydroxyl group and the diol (*minus one hydrogen*).

Next, the proton strikes again! It can be attracted to the hydroxyl group which falls off as water (*a great leaving group*), thus we have a secondary carbocation, again. Now we have a partial negative charge (the oxygen of the free arm of the diol) and a positive charge (the carbocation) in close proximity in the same molecule! In a very fast *intra*molecular reaction, the nucleophile meets the carbon nucleus and regenerates the proton catalyst. The product is answer choice C., a ketal.

Q 2 presents two important points: (i) it describes a compound called LDA as *hindered* (=bulky) and (ii) its conjugate acid has a pK_a equals $-\log K_a$ and K_a is the acid dissociation constant. A K_a which is extremely low ($\approx 10^{-40}$), indicating that the conjugate acid is verrrrry weak, would give a pK_a of about 40. All to say, weak conjugate acid = strong base! Translation of Q2: *what does a strong bulky base like to do?*

Before we answer, what does a strong *small* base like to do in organic chemistry? They tend to engage in nucleophilic substitution reactions (i.e. OH^-, CN^-). A big bulky base usually has difficulty accessing carbon for a nucleophilic reaction. Instead, it takes the easy way out: it plucks protons of a molecule like oranges off a tree!

Of course some H^+'s are easier to remove than others. This depends on the stability of the suggested product. For example, removing a H^+ from the methoxy substituent would create a primary carbanion (verrrrry unstable) with no real stable options to get rid of the negative charge.

On the other hand, examine the carbon in the ring which is attached to the methyl group. If you remove the only hydrogen attached to that 'ringed' carbon (= α-hydrogen which is happy to be plucked, you get a *tertiary* carbanion. Furthermore, you get a logical ensuing reaction: the negative charge is quickly attracted to the carbonyl carbon which is δ^+ thus forming a double bond. Simultaneously, the pi electrons from the carbonyl group are kicked up to the oxygen creating an enolte anion (*almost identical to keto-enol tautomerism.*

Q 3. This question is quick and easy! By looking at Step IX we notice $NaBH_4$, a very important reducing agent. Thus the carbonyl group becomes an alcohol group (*hydroxy*) in a *reduction reaction*.

80

Answers

Q. 5. The correct answer is (C).

Q. 6. The correct answer is (C).

Q. 7. The correct answer is (B).

Q. 8. The correct answer is (C).

Q. 9. The correct answer is (C).

Q. 10. The correct answer is (D).

Answer Explanations

5. $H-C\equiv C-H + HOH \rightarrow H_2C=C-H \rightleftarrows H_3C-C=O$

   with OH on the middle carbon (enol form) and H on the carbonyl carbon (keto form).

6. The weaker acid holds on more tightly to its proton than the stronger acid, which will release its proton more readily. As a result, the concentration of the weaker acid will increase over the stronger acid.

7. The aromatic structure is lost in the keto form so that the enol structure is favored.

8. $-C=C-N-R' \rightleftarrows [-C=C-N-R' \leftrightarrow -C-C=N-R'] + H^+ \rightleftarrows -C-C=N-R'$

9. The imine is the chief product.

10. There is no hydrogen joined to N.

Answer

Q. 11. The correct answer is (C).

Answer Explanation

11. If only $(CH_3)_3CCHO$ and a strong base are present, no reaction can occur. For an aldol condensation to occur, the aldehyde or ketone must contain an α-hydrogen. The α-hydrogen is acidic and will combine with the base to form a carbanion. The carbanion is the nucleophile in the reaction that attacks the carbonyl carbon of another aldehyde or ketone molecule. The generally accepted mechanism follows, using acetone as an example.

$$CH_3-\underset{\underset{O}{\|}}{C}-CH_3 + OH^- \rightleftarrows CH_3-\underset{\underset{O}{\|}}{C}-\overset{..}{\underset{\ominus}{C}H_2} + H_2O$$

$$CH_3-\underset{\underset{O}{\|}}{C}-\overset{..}{\underset{\ominus}{C}H_2} + CH_3-\underset{\underset{O}{\|}}{C}-CH_3 \rightleftarrows CH_3-\underset{\underset{O^\ominus}{|}}{\underset{|}{C}}-CH_2-\underset{\underset{O}{\|}}{C}-CH_3$$
$$ CH_3$$

$$CH_3-\underset{\underset{O^\ominus}{|}}{\underset{CH_3}{C}}-CH_2-\underset{\underset{O}{\|}}{C}-CH_3 + H_2O \rightleftarrows CH_3-\underset{\underset{OH}{|}}{\underset{CH_3}{C}}-CH_2-\underset{\underset{O}{\|}}{C}-CH_3 + OH^-$$

Answers

Q. 11. The correct answer is (B).

Q. 12. The correct answer is (C).

Q. 13. The correct answer is (A).

Q. 14. The correct answer is (C).

Q. 15. The correct answer is (D).

Answer Explanations

11. Step 2 of both mechanisms involves transfer of an aryl group with its bonding electrons to the adjacent carbon atom (carbanion rearrangement).

12. $^-OCH_3$ will add to one carbonyl group forming $C_6H_5-\underset{\underset{C_6H_5-C=O}{|}}{\overset{\overset{O^-}{|}}{C}}-OCH_3$ according to Mechanism A.

13. Oxygen exchange with O^{18}-labeled water will indicate that step 1 is reversible (and faster than step 2).

14. The aliphatic analogs undergo aldol condensations in alkaline solution.

15. Reactions include acylation of the alcohol group and reaction of the carboxyl group (amide formation, esterification).

Answers

Q. 16. The correct answer is (B).

Q. 17. The correct answer is (C).

Q. 18. The correct answer is (D).

Q. 19. The correct answer is (A).

Q. 20. The correct answer is (C).

Answer Explanations

16. In Reaction 1, Cl⁻ adds, while in Reaction 2, the neutral base :NH$_2$OH attacks the β carbon. Attack at the β position produces the intermediate

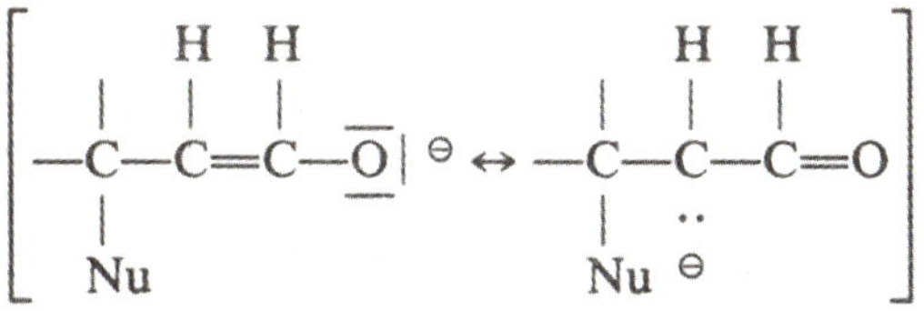

    $$\left[\begin{array}{c} H\ H \\ |\ \ | \\ -C-C=C-\overline{O}| \\ | \\ Nu \end{array} \right]^{\ominus} \leftrightarrow \left[\begin{array}{c} H\ H \\ |\ \ | \\ -C-C-C=O \\ | \\ Nu^{\ominus} \end{array} \right]$$

    which is resonance-stabilized. By adding to an end of the conjugated system, the electronegative oxygen can carry the negative charge.

17. Addition to oxygen yields the carbocation $\left[-C\cdots C\cdots C-OH \right]^{+}$ in which the positive charge is shared among three carbon atoms. Addition to carbon would give the intermediate

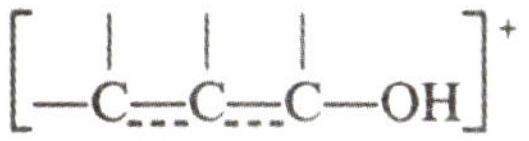

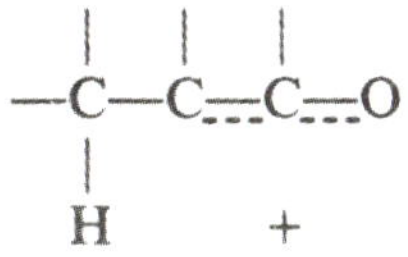

    $$-C-C\cdots C\cdots O$$
    $$|$$
    $$H\ \ \ +$$

    which is less stable.

18. This is a nucleophilic addition of a carbanion to an α, β-unsaturated carbonyl compound. Again, attack occurs at the end of the conjugated system.

20. The carbonyl group activates the carbon-carbon double bond toward electron-rich reactants.

Answer

Q. 21. The correct answer is (B).

Answer Explanation

21. Because both reactants contain acidic α-H's, two different carbanions will form and attack the carbonyl groups in condensation reactions.

Chapter 20

Answers

Q. 1. The correct answer is (B).

Q. 2. The correct answer is (C).

Q. 3. The correct answer is (D).

Answer Explanations

1.

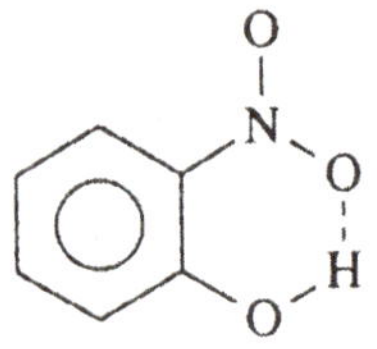

2. Aromatic systems show strong absorptions in the UV region of the system. C—C and C—H bonds will absorb in the IR region. The NMR will show the nonequivalent hydrogens.

3. An equilibrium exists between the enol and keto structures, but it usually lies in favor of the keto form:

   Enol structure ⇌ Keto structure

   Phenol is an exception because the enol form (phenol) is stabilized by the aromatic ring.

Chapter 21

Answer

Q. 1. The correct answer is (C).

Answer Explanation

Q 1 You may recall that a Lewis acid *accepts electrons*. In this way, the catalyst ferric bromide makes the reactant bromine quite positive. Bromine does not like to be positive!!!! Now Br really, really, really wants electrons (= powerful electrophile). That's why it can add to the electron dense benzene ring.

Chapter 22

Answers

Q. 1. The correct answer is (C).

Q. 2. The correct answer is (D).

Q. 3. The correct answer is (B).

Q. 4. The correct answer is (D).

Q. 5. The correct answer is (A).

Answer Explanations

Now that we know each other, we don't have to go into too much detail! The passage is self-explanatory so let's look at the questions. . .

Q 1 shows how irrelevant a passage can be! This question is testing your ability to combine a few important ideas: (i) substituents affect the acidity of phenols; (ii) halides (i.e. Br) are weakly deactivating groups but they are O-P Directors; (iii) activating groups (O-P Directors *except* halides) decrease the acidity of the phenol; (iv) in summary, where EDG = electron donating group and EWG = electron withdrawing group, we get:

EDG	**EWG**	**EWG: Halogens**
activates the ring	*deactivates the ring*	*weakly deactivating*
O/P Directing	Meta Directing	O/P Directing
i.e. alkyl groups	i.e. nitro (-NO$_2$)	i.e. bromine
acid weakening	acid strengthening	acid strengthening
↑ pK_a	↓ pK_a	↓ pK_a

The Reasoning: Electron withdrawing groups can stabilize the *negative* charge on oxygen which encourages oxygen to lose a proton (i.e. *become more acidic*). When a compound becomes more acidic, K_a *increases* thus pK_a *decreases* because pK_a = – log K_a.

Q 2 tests your ability to read the diagram. For each compound, there are only two choices: Sol. (*soluble*) or Insol. (*insoluble*). By following the diagram, step by step, only one compound can be *insoluble* in water, NaOH, HCl, and H$_2$SO$_4$: compound IX. Why not compound III? Because it is *soluble* in NaOH!

Q 3 is seeking to remind you that benzoic acid is a carboxylic acid with 7 carbon atoms. By quickly examining Table 1 in the passage, you may agree that Group III includes benzoic acid. And finally, looking at Group III in Figure 1 makes the answer reveal itself!

Q 4 sends us back to Figure 1. Group I is soluble in water which is defined as *hydrophillic*. The only possible answers are A. and D. The former is not the better answer since it only mentions the *basic* feature of amino acids. But of course amino acids, even small ones, contain both basic and acidic groups = *amphoteric*.

Q 5 can be solved by drawing a quick diagram of the acid-catalyzed ester hydrolysis reaction. For your own interest consider drawing the whole mechanism (in the exam it should quickly fly across your cerebral cortex).

The story goes something like this: The catalyst (H$^+$), being the most charged substance, is implicated first. Thus the electrons from the electronegative oxygen (O*) are attracted to the proton (H$^+$) and bonds. Now oxygen (O*) has the positive charge. Electronegative oxygen does not like to carry positive charges! Thus with haste it draws electrons to itself from a covalent bond (i.e. its bond to the central carbon). Thus HO*R' is the leaving group and now the central carbon carries the positive charge. Now the juicy lone pair of electrons on the oxygen in H$_2$O is quickly attracted (= *nucleophile*) and then bonds to the carbocation. The extra hydrogen in water is kicked out as a proton (regenerating our catalyst) producing a carboxylic acid.

Back to the question: if water is exchanged for an alcohol, the only thing that changes is the last step in the story. Now the alcohol is the nucleophile. Instead of producing a carboxylic acid, we produce another ester.

Answers

Q. 6. The correct answer is (B).

Q. 7. The correct answer is (C).

Q. 8. The correct answer is (C).

Q. 9. The correct answer is (C).

Q. 10. The correct answer is (A).

Q. 11. The correct answer is (B).

Answer Explanations

Q 6 can only go wrong if you're "thinking too much"!! On the most basic level, in the first reaction we saw a large ugly molecule on the left side of the equation and a *similarly* ugly molecule on the right side. The difference being 2 H's and 1 O. Mmmmm, 2 H's and 1 O, H$_2$O, water! Superkeeners will complain about the imperise way the reaction is written: *where did the second molecule come from?* When you're offered a gift, don't reject it because it's not exactly the way you wanted it to be!

Q 7 is trying to test your *understanding* of a mechanism. This is done by labelling oxygen in some way and seeing if you can work out where *that* oxygen, in water, ends up in the product. First we need to determine which reaction involves water as a reactant. A quick look at Figure 1 reveals H$_2$O on the third line. Of course, oxygen in H$_2$O is δ$^-$ and thus will be attracted to a δ$^+$ atom. The carbon double bonded to the nitrogen must be δ$^+$. Furthermore, since nitrogen is δ$^-$ and the hydrogens in H$_2$O are δ$^+$ then their mating is also inevitable. In summary, the hydrogens in H$_2$O help form the amino group –NH$_2$ while the oxygen in H$_2$O, which we are tracing, helps the second carbonyl group, C=$\underline{O}$, in the illustrated product.

Q 8 tests your recognition of the δ$^-$ nitrogen in the amino group as a nucleophile (= *nucleus-loving, AKA positive charge-loving!*). Now only answer choices B. and C. are possible. Recall that nitrogen's lone pair electrons are essential to carrying out a nucleophilic reaction. In the reaction provided in the question, we go from a nitrogen with four bonds (= *no lone pair available*) to one with three bonds thus the addition of a base freed up the lone pair which can now engage in a nucleophilic reaction. Furthermore, the free amino group prior to the addition of a base was positively charged thereby being a distinctly terrible nucleophile!!

Q 9 begins with the issue of nomenclature. Alanine should be recognized as an amino acid. Draw the *general* structure of an amino acid below.

Benzoyl chloride should be recognized as an acid halide. Draw the general structure of an acid chloride below; *note that 'Benz' refers to benzyl, as in the benzene ring, which is stable to nucleophilic reactions so we need not account for its presence!*).

Now if you were able to get Q8 correct then Q9 is just a follow-up. The *amino* acid contains a δ^- nitrogen in the *amino* group which is a nucleophile. The nucleophile is attracted to the δ^+ carbon of the carbonyl group in benzoyl chloride. The weakest bond breaks (*the π bond within the double bond*), and the electron pair lands on the electronegative oxygen. Now carbon is surrounded by *three* electronegative atoms making carbon verrrrry δ^+! This is an unstable situation!

Thus the negatively charged free electrons on oxygen quickly mate with the verrrrry δ^+ carbon. However, since carbon can only be bonded four times, one of its substituents has to go! Since chloride is an excellent leaving group, we are left with an *amide*. Draw the mechanism below.

Why was the medium "dilute aqueous sodium hydroxide"? Once again we turn our attention back to Q8. Base treatment of an amino acid increases the rate of a nucleophilic reaction of the free amino group.

Would the "dilute aqueous sodium hydroxide" create other products in the reaction? Because it is *dilute* we would expect very little contamination or by-products. But just for fun, let's treat both our reactants with *concentrated* aqueous sodium hydroxide.

Now NaOH in water (= *aqueous*) means Na$^+$ and OH$^-$, which simply means the nucleophile OH$^-$ since Na$^+$ is a spectator ion. Note that both compounds have a carbonyl group which are internationally popular in attracting nucleophiles! Treat your preceding reactants with OH$^-$. The nucleophile is quickly attracted to the δ^+ carbon of the carbonyl group. The weakest bond breaks (*the π bond within the double bond, ORG 1.3.1*), and the electron pair lands on the electronegative oxygen. Now carbon is surrounded by *three* electronegative atoms making carbon verrrrry δ^+! This is an unstable situation! {*does this sound familiar?!*}

Thus the negatively charged free electrons on oxygen quickly mate with the verrrrry δ^+ carbon. However, since carbon can only be bonded four times, one of its substituents has to go! In the case of the amino acid, the hydroxyl group leaves simply re-establishing the carboxylic acid. In the case of the acid chloride, chloride is an excellent leaving group, thereby creating a *carboxylic acid*.

Q 10 reminds you of Principle #4: isn't it remarkable how many questions in a row that you can answer without reading the passage?

This question can be answered very quickly with some strategy and some basic info about amino acids. Amino acid I. has the side group C_6H_5- which is the benzene ring which is neither *ionic* nor *polar* and is thus *hydrophobic*. Amino acid II. has the side group –OH which *is* polar and thus *hydrophilic*. Therefore, without going any further, the answer must be A.

For interest, we note that amino acid III. has the amino side group which is protonated at neutral pH (= *positively charged*). Amino acid IV. has the carboxylic *acid* side group which *loses a proton* at neutral pH (= *negatively charged*).

Q 11 is really asking: what is the better base (*proton acceptor*) an amino group –NH$_2$ or the carboxylate anion –CO$_2^-$, which is an anion because carboxyl is an *ACID* (*carboxylic acid*). Mmmmm, I'd go with the amino group also!

Answers

Q. 12. The correct answer is (B).

Q. 13. The correct answer is (A).

Q. 14. The correct answer is (B).

Q. 15. The correct answer is (D).

Q. 16. The correct answer is (C).

Answer Explanations

Q 12 can be answered without looking at any of the information in the problem! By definition, alkanes (i.e. Heptane) are hydrophobic (= *'insoluble in'* or *'afraid of' water*). The other answers are nonsense.

Q 13 tests your ability to use the figure and table. Follow the flow chart in Figure 1. Water insoluble + 10% NaOH soluble + 5% $NaHCO_3$ soluble = a carboxylic acid (see the preceding example we worked through). Then we try 2% $KMnO_4$ (*potassium permanganate*), which table 1 tells us is the Bayer test. According to Table 1, the Baeyer test is positive for alkenes. According to Q13, the Baeyer test is negative. Thus we are looking for a carboxylic acid (–COOH) which is not an alkene. 'A.', which is benzoic acid, is the only possible answer.

You must be comfortable with these types of problems because flow diagrams are MCAT classics!

Let's work through an example. Consider Figure 1. The flow chart begins with an unknown compound which is not soluble in water. The compound is then treated with 10% NaOH. If the compound is insoluble in the NaOH then we test the compound in 2,4-DNP (*of course, the mechanism of all these reactions and what 'DNP' stands for are irrelevant!*). Had the compound been soluble in NaOH then Table 1 tells us that the unknown compound must be an organic acid (= *a carboxylic acid* or *a phenol*). To determine what type of organic acid, Figure 1 shows us that the next step is treatment with 5% sodium bicarbonate, $NaHCO_3$. If the test is positive (i.e. the organic acid is soluble in $NaHCO_3$), the Table 1 tells us the compound must be a carboxylic acid.

Q 14 is asking you to remember the following simple facts: (i) signals on a proton NMR are used to determine the *number* and *types* of *hydrogens* in a molecule; (ii) an *acid* can donate a proton (hydrogen); (iii) if the solvent is D_2O instead of H_2O then the *acidic proton* can be replaced be deuterium instead of another proton; (iv) loss of a proton means loss of a signal.

Q 15 is really asking the following question: what determines basicity in an amine? And an even more basic question (*pardon the pun!*), what is a base?

A base accepts protons. This means a base must possess a negative or partially negative charge in order to attract a proton. The partial negative charge on an amine is carried by nitrogen. Clearly, substituents attached to the nitrogen which donate electrons will increase the partial negative charge on N thus increasing its attractiveness to a proton which makes the amine a better base. Conversely, substituents which withdraw electrons from N will decrease its partial negative charge which decreases its attractiveness to a proton which means decreased basicity.

Now the problem is quite simple. The differences between the three aliphatic amines are: I. has no electronegative substituent; II. has fluorine which is *the most electronegative element in the periodic table;* and III. has chlorine which is very much electronegative but not as much as fluorine. Thus the order of basicity is: I., the most basic, then III. with chlorine helping to withdraw electrons away from N, then II., the weakest base of the three because the electronegative strength of F pulls electrons from N the most.

Q 16 is very, very, very, very easy! The passage tells you and we worked out that a positive $NaHCO_3$ test indicates a carboxylic acid. Answer choice C. is the only choice which is a carboxylic acid! Everything else is irrelevant!

{*Please note how many questions in the preceding passage could have been answered without reading the passage! Though this is certainly not always the case, it does underline an important point: if you are running out of time in any science section on the MCAT, skip reading the passages and go straight to the questions.*}

Answer

Q. 17. The correct answer is (D).

Chapter 23

Answers

Q. 1. The correct answer is (B).

Q. 2. The correct answer is (B).

Q. 3. The correct answer is (A).

Q. 4. The correct answer is (C).

Q. 5. The correct answer is (D).

Answer Explanations

Prelude: you don't need to understand every detail in order to answer these questions correctly. Nonetheless, let's pull out the magnifying glass . . .

Begin by reading the first paragraph. The key words are 'conjugated dienes' and 'electrophilic addition'. A *diene* is an alkene with two double bonds (i.e. 1,3-buta*diene*; the numbers 1,3 locate the double bonds, but- means 4 carbons, see Reaction I in the problem). *Conjugated* means there is a single bond between the two double bonds. *Electrophilic addition* is the most important mechanism for you to understand regarding alkenes!

Consider the second paragraph and Reaction I. Let's discuss the mechanism. First we need to understand the nature of a double bond: *it is an electron dense area; therefore, the electrons attract positively charged substances* = electrophiles. A classic electrophile is H^+. Thus in a prototypical electrophilic addition reaction a compound like HBr is added to an alkene. The HBr ionizes and the electrophile H^+ adds to the double bond in accordance to Markovnikoff's rule. For short we'll call it Mark's rule! Some students like to memorize Mark's rule as follows: the electrophile or H^+ prefers to add to the carbon where most of its buddies (other H's) are! Then the nucleophile (i.e. Br^-) attaches to the carbocation which is *usually* more substituted (i.e. more attached groups which does not include H's); these groups, i.e. alkyl groups, are somewhat electron donating and thus stabilize the carbocation intermediate.

Umm, one question: where the heck is the electrophile in Br_2? After all, it's a diatomic molecule where both atoms are identical. For this we must revert to an analogy . . .

Consider two 5 year old girls, identical twins, walking in a park hand in hand with rather *neutral* facial expressions. Someone pulls out an ice cream cone which is presented to the closer of the two twins. The twin closest to the ice cream begins to smile (a *positive* facial expression), while the other twin realizing she is going to miss out on the treat begins to frown (a rather *negative* expression). An outside force has *induced* polarity in the twins = *an induced dipole!*

When Br_2 approaches a double bond, the polarity changes. The electrons in the double bond repel the electrons in the bond between the bromine atoms making the distribution of electrons among the 'twins' unequal. Thus one Br atom becomes slightly negative (δ^-), while the other becomes slightly positive (δ^+). The δ^+ Br is the electrophile which becomes increasingly attracted to the double bond which in turn continues to repel the Br-Br bond until the electrophile Br attaches while the leaving group Br^- floats away before acting as a nucleophile.

Back to Reaction I: we now know how to create the intermediate. The double bond induces a dipole in Br_2 resulting in one Br atom adding to the carbon with the most H's (*Mark's rule*), leaving a carbocation as an intermediate. Draw the reaction below:

Q 1 B is the only possible answer since it is the only answer with carbocations as intermediates! You should have them in your reaction above.

Q 2 is very nice! It tests your understanding of the following concept: the more stable intermediate requires the *smallest* activation energy and thus forms a product *faster* (= *the kinetic product*, A), while the less stable intermediate requires *more* energy (*higher temperature*, i.e. 40 °C in the problem) to form a product (= *thermodynamic product*, B).

Q 3 is a continuation of the preceding paragraph. The real question is: which intermediate is the most stable? By looking at the two intermediates (Q1, II.), you can see that the compound on the left has a positive charge on a 2°carbon (= *secondary carbocation*), while the compound on the right has a positive charge at the 1° carbon (= *primary carbocation*). Since the secondary carbocation is more stable, it must produce compound A. Now simply imagine our nucleophile Br⁻ adding to our secondary carbocation intermediate. Now it's just a matter of nomenclature.

Q 4 is a direct definition of a structural isomer.

Q 5 is also quick and easy! To begin with, answer choice B. is incorrect because it's a *five* carbon ring while our reactant is a *six* carbon ring (cyclo*hex*ene). Answer choice C. also is illogical since electrophilic addition adds to the double bond! For fun, let's name answer choices A. and D.: A. is *cis*-1,2-dibromocyclohexane (*cis* tells us the two Br atoms are on the same side of the plane of the cyclohexane molecule), and D. is *trans*-1,2-dibromocyclohexane (*trans* tells us the two Br atoms are on opposite sides of the plane of the cyclohexane molecule). Which is more stable? Naturally the electronegative Br atoms would prefer to be as far apart as possible because of *electron shell repulsion*. The *trans* molecule affords maximum distance between the Br atoms and is thus more stable.

Answers

Q. 6. The correct answer is (D).

Q. 7. The correct answer is (A).

Q. 8. The correct answer is (C).

Q. 9. The correct answer is (A).

Q. 10. The correct answer is (C).

Answer Explanations

6. Experiment 2 shows that 1,3-butadiene reacts more rapidly than 1-octene.

7. These two structures account for the 1,2- and 1,4-addition products shown in step (4).

8. The magnitude of the activation energy determines the rate of the reaction. Activation energy is the *difference* in energy between the reactant(s) and the transition state.

9. Free Radical• $CH_2 = CH - CH = CH_2$
   $CH_2 = CH - CH = CH_2$
   $[-CH_2 - CH = CH - CH_2-]_n$

10. As with free radical addition, the 1,2- and 1,4-addition products are formed.

Chapter 25

Answer

Q. 1. The correct answer is (C).

Answer Explanation

Q 1 This question tests your memory of the direction of bonds. This question should be easy. The structure of glucose has six carbons (*this should be consistent with your knowledge of the molecule!*). Only answer choices A. and C. have six carbons. The structure given never has 2 hydroxy groups attached to the same carbon. That eliminates A. as an option.

Answer

Q. 2. The correct answer is (C).

Answer Explanation

Q 2 This can be solved quickly: focus on the –CH_2OH substituent in Figure 1. The shortest route between it and the circled hydroxyl contains *three* carbons. Now look at the –CH_2OH substituent in the modified Fischer projection (Fig. 2) of β-D-glucose. We should be able to count three carbons along the shortest route between it and I, II, III, or IV. The answer must be III.

Answer

Q. 3. The correct answer is (D).

Answer Explanation

Q 3 The glucose anomers are diastereomers that differ in configuration about C—1.

Answers

Q. 4. The correct answer is (D).

Q. 5. The correct answer is (C).

Q. 6. The correct answer is (D).

Q. 7. The correct answer is (B).

Q. 8. The correct answer is (D).

Q. 9. The correct answer is (B).

Answer Explanations

4. D-(+)- Glucose does not undergo some reactions typical of aldehydes: Although it is readily oxidized it does not form a bisulfite addition product and it gives a negative Schiff test. Also, when D-(+)- glucose (mp = 146°C) is dissolved in water, the specific rotation drops from 112° to 52.7° while crystals (mp = 150°C) are dissolved in water, specific rotation rises from 19° to 52.7°. α-D-(+) Glucose is the form with the higher positive rotation while β-D-(+) glucose is the form with the lower rotation. When each form is dissolved in water, the specific rotation will change to the equilibrium value (mutarotation).

5. When D-(+)- glucose reacts with methanol and HCl, methyl D-glucoside forms in which only one –CH_3 group is added; however its properties resemble a full acetal. Only the structures below are consistent with this data.

   Methyl β–D-glucoside: Methyl α–D-glucoside:

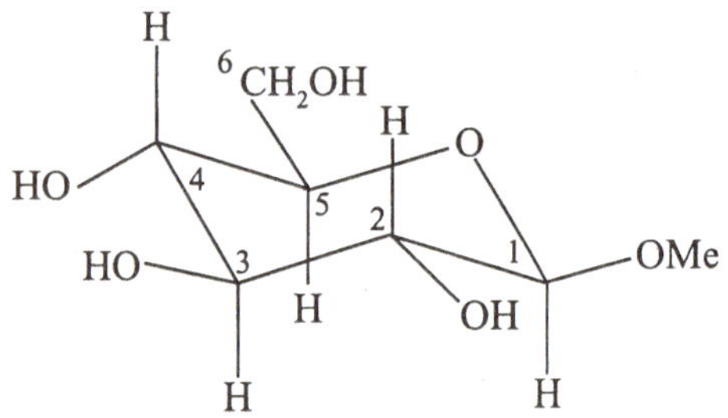

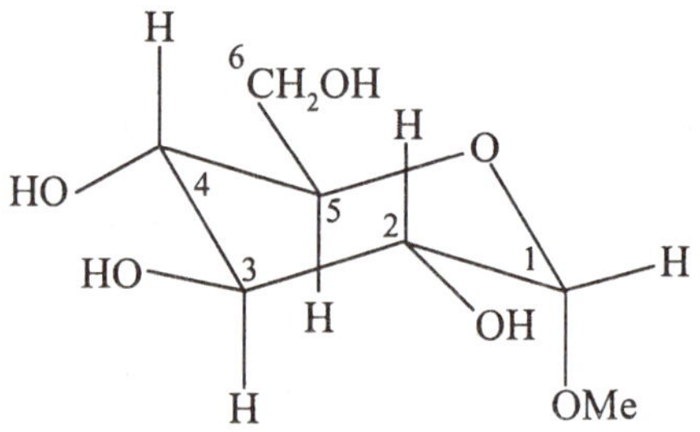

6. α–D-glucose: 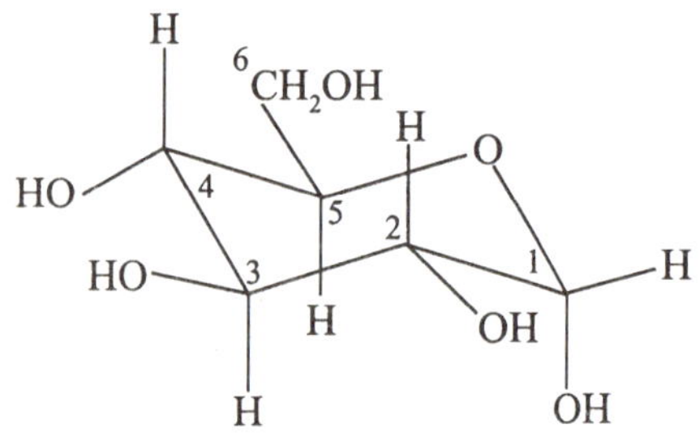   β–D-glucose:

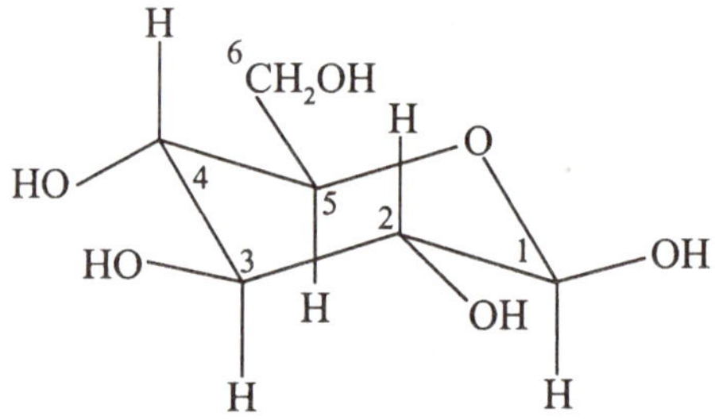

7. See the answer to number 6.

8. Both cellulose and starch consist of chains of D-glucose units, each unit connected by a glycoside linkage to C-4 of the next:

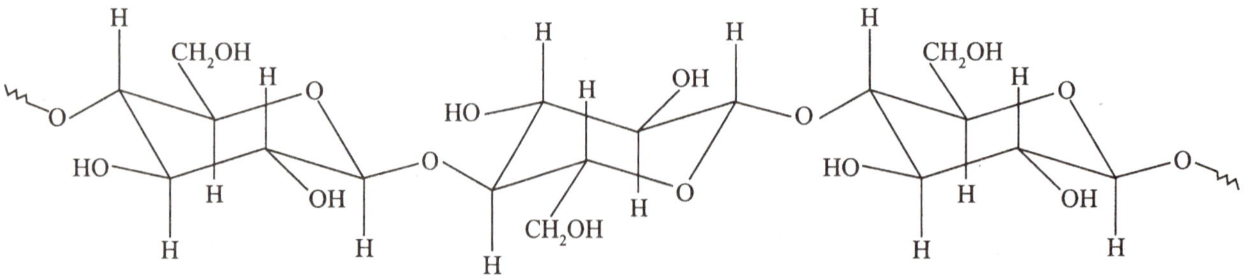

9. Epimers are a pair of diastereomeric aldoses that differ only in configuration about C-2. Epimers have the same configuration about C-3, C-4, and C-5.

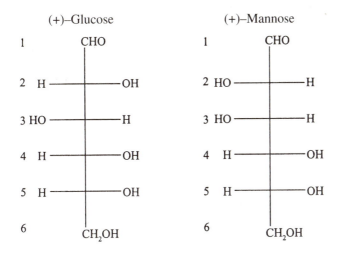

Chapter 26

Answers

Q. 1. The correct answer is (D).

Q. 2. The correct answer is (A).

Answer Explanations

1. Glycerol: a fat is a triglyceride.

```
        O
        ‖
H₂C—O—C—R
|       O
|       ‖
HC—O—C—R'
|       O
|       ‖
H₂C—O—C—R"
```

2. Fats are triacylglycerols:

```
CH₂–O–C–R
|    ‖
|    O
CH–O–C–R
|    ‖
|    O
CH₂–O–C–R
     ‖
     O
```

93

All three ester linkages are to acyl groups. Phosphoglycerides contain only two acyl groups: A phosphate group replaces the third acyl group:

A Phosphogyceride

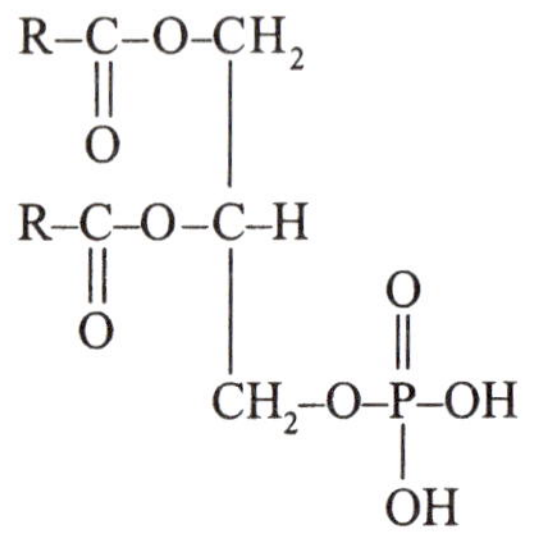

$$\begin{array}{l} R-\underset{\underset{O}{\|}}{C}-O-CH_2 \\ R-\underset{\underset{O}{\|}}{C}-O-C-H \\ CH_2-O-\underset{\underset{OH}{|}}{\overset{\overset{O}{\|}}{P}}-OH \end{array}$$

Chapter 27

Answer

Q. 1. The correct answer is (A).

Answer Explanation

Q 1 You should be familiar with the concept of isoelectric point from the organic chemistry review.

Answers

Q. 2. The correct answer is (B).

Q. 3. The correct answer is (D).

Q. 4. The correct answer is (A).

Q. 5. The correct answer is (C).

Q. 6. The correct answer is (C).

Q. 7. The correct answer is (B).

Answer Explanations

Q 2 There are two equivalence points (or full neutralization points) ii and iv indicating two acidic protons.

Q 3 When the amino acid is acidified its structure is:

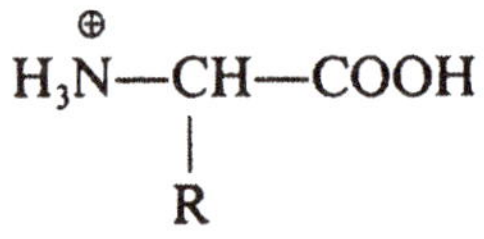

$$H_3\overset{\oplus}{N}-\underset{\underset{R}{|}}{CH}-COOH$$

It therefore behaves like a diprotic acid with two acid dissociation constants, K_{a1} and K_{a2}. The pK_a values are:

$$pK_{a1} = -\log K_{a1}$$
$$pK_{a2} = -\log K_{a2}$$

Each amino acid has its characteristic pK_a values at each 1/2 neutralization or equivalence point (after adding 0.50 mol and 1.50 mol of NaOH pH = pK_a). Therefore, by measuring the pH after addition of 0.50 mol of NaOH you have the value of pK_{a1}. After the addition of 1.50 mol of NaOH, pH = pK_{a2}.

Q 4 $pI = \dfrac{pK_{a1} + pK_{a2}}{2}$ at this pH the amino acid is a zwitterion

$$\overset{\oplus}{H_3N}-\underset{R}{\underset{|}{CH}}-COO^{\ominus}$$

$$pI = \dfrac{2.40 + 9.80}{2} = 6.10$$

Q 5 At this point in the titration the first 1/2 equivalence point has been passed. At the first 1/2 equivalence point one-half of the diprotic acid

$$\overset{\oplus}{H_3N}-\underset{R}{\underset{|}{CH}}-COOH$$

has been converted into its conjugate base

$$\overset{\oplus}{H_3N}-\underset{R}{\underset{|}{CH}}-COO^{\ominus}$$

$$\overset{\oplus}{H_3N}-\underset{R}{\underset{|}{CH}}-COOH + OH^{\ominus} \rightarrow \overset{\oplus}{H_3N}-\underset{R}{\underset{|}{CH}}-COO^{\ominus}$$

Therefore, at the first 1/2 equivalence point

$$[\overset{\oplus}{H_3N}-\underset{R}{\underset{|}{CH}}-COOH] = [\overset{\oplus}{H_3N}-\underset{R}{\underset{|}{CH}}-COO^{\ominus}]$$

As you add more NaOH you increase

$$[\overset{\oplus}{H_3N}-\underset{R}{\underset{|}{CH}}-COO^{\ominus}]$$

so that it exceeds the concentration of the diprotic form.

Q 7 After the addition of 1.5 mol of NaOH

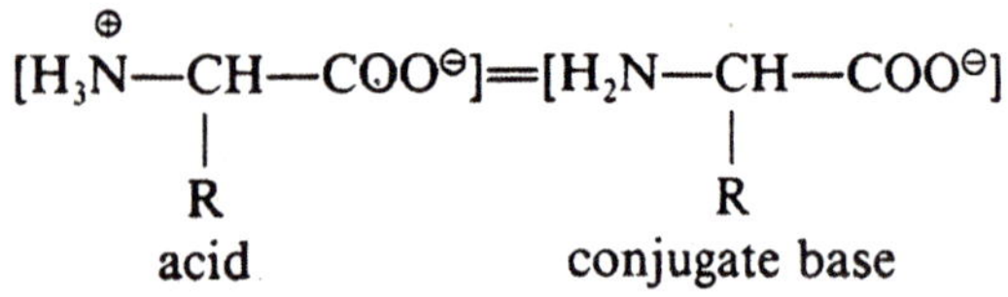

$$[\overset{\oplus}{H_3N}-\underset{R}{CH}-COO^{\ominus}] = [H_2N-\underset{R}{CH}-COO^{\ominus}]$$

acid conjugate base

A buffer solution contains similar concentrations of an acid and its conjugate base (or a base and conjugate acid).

Answers

Q. 8. The correct answer is (D).

Q. 9. The correct answer is (B).

Answer Explanations

8. The amino acid residues are joined by amide linkages:

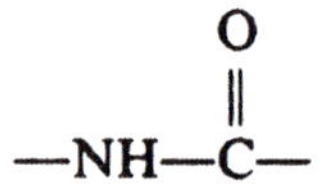

$$-NH-\underset{\underset{}{}}{\overset{O}{\underset{\|}{C}}}-$$

9. Tertiary amines do not react with acid chlorides to form amides because tertiary amines cannot lose a proton after attacking the carbon. The alkylation of amines by alkyl halides is a nucleophilic substitution reaction in which the organic halide is attacked by the nucleophilic amine. Peptides are formed by the reaction between the carboxyl group of an amino acid and the amino terminal of another amino acid to form an amide linkage.

ORGANIC OWL

Please follow these simple steps

1. Go to http://owl.thomsonlearning.com

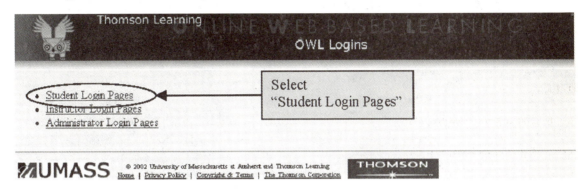

2. Find your textbook

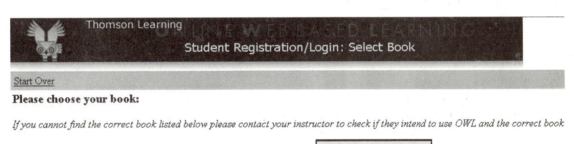

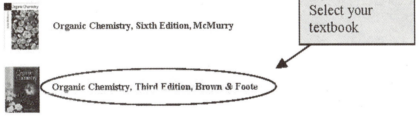

3. find your institution

Thomson Learning
ONLINE WEB-BASED LEARNING
Student Registration/Login: Select Institution

Start Over

Book: **Organic Chemistry, Third Edition, Brown & Foote**

Please choose your institution:

If you cannot find your institution listed below please contact your instructor to check if they intend to use OW.

Quick Link Index:

A A B C D E F G H I J K L M N O P Q R S T U V W X Y Z

Alabama A&M - Normal, Alabama

A A B C D E F G H I J K L M N O P Q R S T U V W X Y Z

Brooks Cole - Thomson Learning, Belmont, California

A A B C D E F G H I J K L M N O P Q R S T U V W X Y Z

Hartwick College - Oneonta, New York
Hofstra University - Hempstead, New York

Institutions are in alphabetical order. Select by scrolling down the list or selecting the letter your institutions name begins with.

4. Find your department

Thomson Learning
ONLINE WEB-BASED LEARNING
Student Registration/Login: Select Department

Start Over

Book: **Organic Chemistry, Third Edition, Brown & Foote**
Institution: **Brooks Cole - Thomson Learning, Belmont, California**

Please choose your department:

- Choose *Student Login Page* if you have already registered for OWL in the department.
- Choose *OWL Registration* if you have not registered yet.

If you cannot find the correct department listed below please contact your instructor to check if they intend and what department to choose.

Department	Student Login Page	OWL Registration
Organic Chemistry	→	→

Select "OWL Registration" by clicking the red arrow.

UMASS © 2003 University of Massachusetts at Amherst and Thomson Learning
Home | Privacy Policy | Copyright & Terms | The Thomson Corporation
THOMSON

5. Find your section #

6. Follow instructions for completing registration form.

ACCESS CODE: b0502-2p1fr-2p2yj-vpaqv-uhr53